Turning Kids On To SCiENCE

IN THE HOME

BOOK 4

LIVING THINGS

Easy to follow recipes for doing 50 science activities
on Plants and Human Biology carried out
with very simple materials
found in and around the house.

Dr. Tik L. Liem

"TURNING KIDS ON TO SCIENCE in the home
 Book 4 - LIVING THINGS"

has been published by:

SCIENCE INQUIRY ENTERPRISES
14358 VillageView Lane
Chino Hills, California 91709

Copyright ©1992 by **Tik L. Liem**

ISBN: 1-878106-07-4 (Volume 4)
ISBN: 1-878106-08-2 (4 Volume set)

Printed in the United States of America

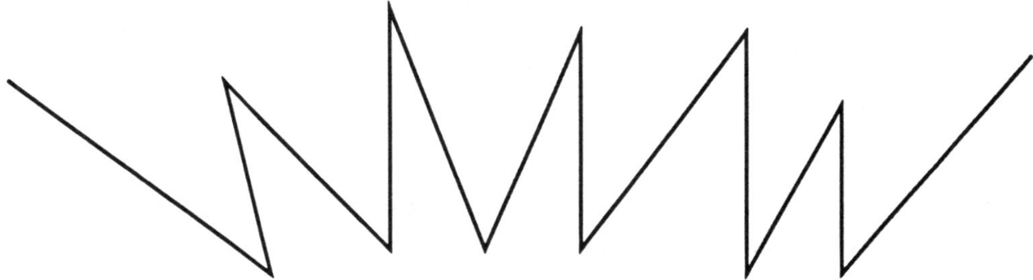

TO THOSE PARENTS
WHO WANT TO PUT MORE *Excitement*
IN THEIR CHILDREN'S LIFE

* Only with adult supervision may children use this book!

* Fire may only be used with adult supervision!
  All fire hazardous events are marked with this symbol:

* Chemicals and/or electricity may only be used under adult supervision!
  Never touch chemicals with your bare fingers: poisonous chemicals
  can even be absorbed through the skin!
  Hazardous events involving chemicals are all marked with: 

* The use of chemicals always brings with it some hazards. It can be
  dangerous, but when handled properly it can be lots of fun!

* When working with a flame or when a flame is involved in an event,
  always have a carbondioxide or Halon fire extinguisher at hand.

* Always follow instructions - the "How to do it" - very carefully!

* Never work all by yourself. Always have someone with you ready
  to turn the fire or electricity off, if something goes wrong.

* The author of **"Turning Kids On To Science In The Home"** is not
  responsible for any accidents resulting from irresponsible or incom-
  petent use of the demonstrations and activities suggested in this book.

* Although some events are rather dangerous, but if the instructions
  are carefully followed, they can give you a lot of joyful moments!

So, have lots of fun!

CONTENTS BOOK 4

TO BE *Enthusiastic*

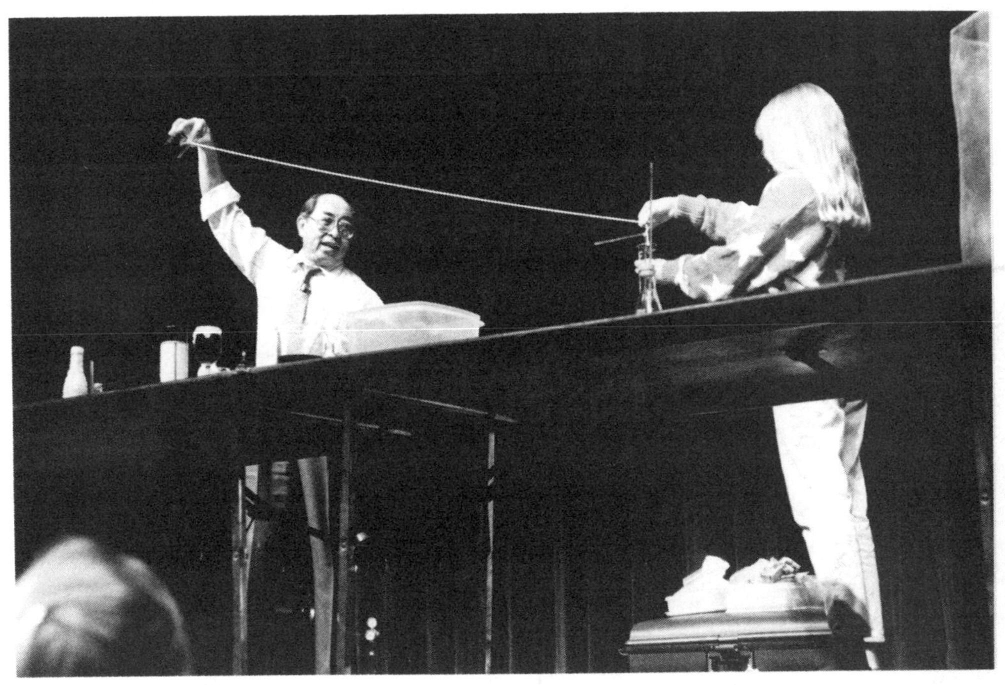

ACT *Enthusiastic*

AND YOU WILL BE !

INTRODUCTION to BOOK 4

This book is the fourth book to a sequence of four volumes, written and designed for parents of students of science, particularly for those at the lower and upper elementary, and junior high or intermediate level, for senior high students, for college students preparing to teach science, and for all those individuals who are interested in science and the application of science in our daily life.

In this space age where science plays a major role in determining whether a country or even a personal life will be able to advance or progress, I feel that the understanding of science concepts and science principles will strongly help towards this goal. Statistics and test results have shown that American youngsters can't perform the simplest technical functions. A recent report of a study done by the Educational Testing Service in Princeton, New Jersey mentioned that American nine-graders were compared with same age level children of five other countries and four Canadian provinces, and they came out on the bottom in mathematics and very near the bottom in science! This is very sad news, even worse is their prediction that in the year 2000 there will be a shortage of close to a quarter million engineers, physicists, chemists, and biologists in the USA.

Where lies the problem? What makes the USA so lagging behind other countries, like: Japan, Western European countries, and in recent years some of the Pacific Rim countries? This lagging is not only economically, but especially in academic achievements of students in the field of mathematics and science! Could it lie in the way math and science is being taught in our schools? Could it lie in the way math and science is being presented to our young children? These questions come up in my mind when too many children I talked to said, that they do not like math and science in their schools. Could the problem be that school teachers do not have enough background in mathematics and science? And because of this, that they are conveying the message to students that math and science are a chore and really not fun to study? Could it be that schools do not have enough materials for doing science? Could it be that at home, our young children are also not getting any encouragement from their parents to go in the fields of science and mathematics?

I think it is a combination of all those mentioned, especially I think that not enough **enthusiasm** in science is being shown in the home as well as in the schools. If teachers as well as parents of students can show more excitement and **enthusiasm in science**, and not be afraid to actually show their enjoyment in doing science in front of the children, there is no doubt in my mind that more of our younger people will find science exciting, and continue studying in the field of science!

In the teaching of a science concept, it is most essential for the teacher to arouse the students' curiosity. Unless the student wants to know what the teacher has to say, it is most likely that time and effort in trying to teach the student would have been completely wasted. The first task of a teacher is to attract the attention of the student. This goes for all levels of learning, whether it is primary grade or college. A student without interest is the desperation of any teacher. In fact, the main objective of formal education is the arousing of the students' interest, if nothing else. Once curiosity is aroused the student will learn much more on his/her own than the teacher can ever teach him/her. Learning only takes place if the student wants to learn!

In this book the description of each of the **exciting events** is organized in such a way as to provide the user with the maximum guidance for conducting a science inquiry lesson. Emphasis has been placed on the use of simple material so that most of the events may be carried out with things that are found in everyday life or may be bought in local stores.

When you as a parent, use this book, you are actually becoming your child's or children's own private teacher! Isn't that exciting! All the statements concerning the teacher apply therefore, to you as well. The questions are there for each of the **exciting events** to be posed to the learner. The explanation is always there on each of the pages to give you the support you need in answering the questions. You can, of course, add your own questions if you want to go into more depth in each of the events.

The use of exciting events in the teaching of science is one of the best methods to arouse interest and curiosity. These **exciting events**, sometimes also called **discrepant events**, mind-capturing, or intuition-offending, are occurrences or happenings which go against what we usually think likely. Each of the events is set up in such a way that it poses a question to the student and asks him/her to come up with the explanation. There is evidence that students will remember science concepts longer when they were engaged in experiencing **exciting events**. This series of books is a collection of thoroughly tested **exciting events**. They can be used to start a lesson in almost any topic of science at the elementary or secondary level. They can also be used as reinforcement activities in the homes or as challenging problems for further inquiry. Each of the events is a focus for inquiry teaching, a most appropriate way of reflecting the true nature of science in the homes. The students are required to put into practice most of the science processes when engaged in inquiry.

The science concepts in this series of books are organized in such a way that it makes most sense:
Book 1: **Our Environment**, consisting of the chapters: Air, Flowing Air, Weather,
           Characteristics of Matter, and Chemicals in our Environment.
Book 2: **Energy**, consisting of the chapters: Forms of Energy, Heat, Magnetism, Static &
           Current Electricity, Light, and Sound.
Book 3: **Forces & Motion**, containing chapters: Forces Affecting Things, Forces and Principles
           in Space Science, Some Phenomena in Earth Science.
Book 4: **Living Things**, containing the chapters: Plants, and Human Biology.

In order to make it easier to find where certain topics are being dealt with, each of the books has its own **table of contents** in the front and an **index** in the back of the book. **Introductions** are placed in front of each book, so that each of the books can be used independently from each other.

INSTRUCTIONS FOR PARENTS

Please follow these instructions for the Teaching Procedure in using **Turning Kids On To Science in the Home** as close as possible, so that your lesson will be most effective:

1. Presentation :

Present to the learner or involve the learner with the **exciting event** by describing or commenting on the names of objects and operations only and <u>not</u> mentioning the reasons for the occurrence. In other words, you may tell the student what you are doing and what materials you are handling, but <u>not</u> the reason why something is happening.

2. Interaction:

Ask the learner questions that eventually will lead him/her to the main reason for the occurrence. In doing this, the students will be engaged in science inquiry and actually practicing the science processes of observing, measuring, inferring, predicting, interpreting data, identifying and controlling variables, hypothesizing, and experimenting.

3. Involvement:

Participate the learner in similar (and simpler) **exciting events** or exciting activities, that are based on and illustrating the same science concept. This will reinforce the learning and retention of that particular concept. Students may work individually or together with you.

It is essential that the presentation of the **exciting event** or the involvement of the students in the event takes place under the following conditions. Think of giving the student: **I N S I G H T !**

IN = Arouse <u>Interest</u>

The event should confront the student with a perplexing problem. It should be presented almost like the way a magician would do a magic act.

S I = Use <u>Simple</u> Materials

The event should be performed with materials that are familiar to the student, in other words use simple materials that can be found in daily life.

G = Use all <u>Gateways</u>

Use all the gateways into the brain, meaning students should have the opportunity to observe as well as carry out the events themselves. They should be allowed to totally experience the exciting event (by using all five senses).

H = <u>Hinge</u> it with Examples

When dealing with the concept underlying the event, examples and applications of the concept in our daily life should be mentioned to make it more meaningful to the student.

T = <u>Tie</u> it all together with Joy and Enthusiasm

Show genuine enthusiasm in presenting the puzzling event and don't be afraid to show your own enjoyment in the subject matter in general!

ENJOY YOURSELVES !

BOOK 4

This volume consists of two chapters: one dealing with plants and the different variables that are influencing their growth, and the last chapter contains demonstrations and activities about human biology.

Chapter 1 deals with the variables affecting the growth of a plant, like: light, gravity, the amount of water. Osmosis and capillary action, respiration and photosynthesis are also dealt with.

Chapter 2 contains demonstrations and activities to start off lessons about the senses, the nervous system, the circulatory, the respiratory, and digestive system in the human body.

THINK POSITIVE !

IF YOU THINK YOU CAN,

YOU CAN !

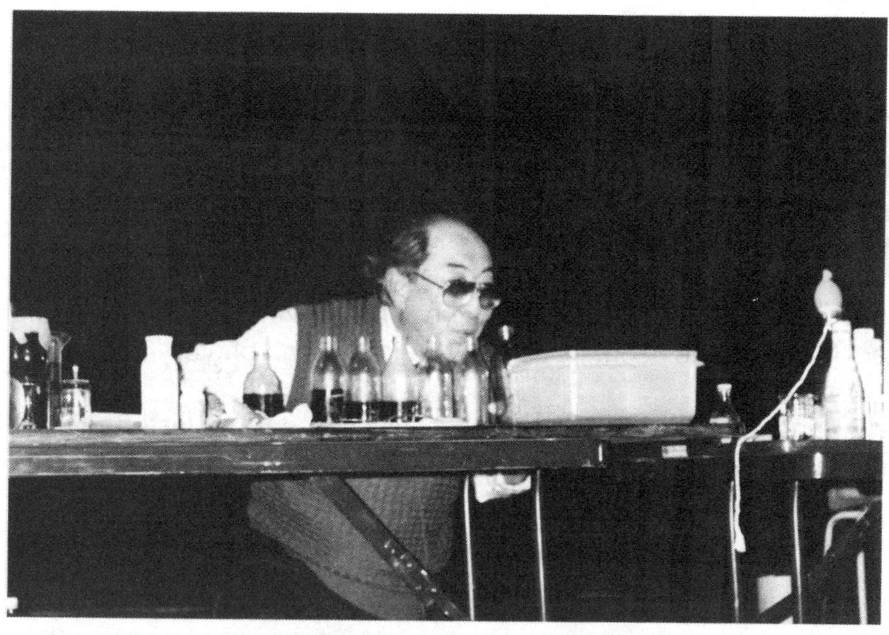

THE REVERSING PITCH

Showing that musical talent can be combined with teaching science !

CHAPTER 1

WHAT VARIABLES ARE AFFECTING THE GROWTH OF PLANTS ?

OBJECTIVES

After dealing with and studying the concepts and sub-concepts in this chapter, the students should be able to:

a. Recognize the correct explanation of an observed event based on each of the sub-concepts;
b. Explain in their own words which of the sub-concepts is determining the course of an event;
c. Distinguish true from false statements concerning each one of the sub-concepts;
d. Identify the correct explanation of an event in daily life applying one of the sub-concepts;

all in relation to the following sub-concepts:

— Conditions for germination of seeds are: moisture, warm temperatures, and air supply.
— Plants grow in the direction of the light source.
— Plant roots grow down towards the center of the earth.
— The parts of a plant that are above the ground grow vertically up.
— Water moves into the plant roots by osmosis.
— Water moves into higher parts of the plant by capillary action.
— Leaves give off water vapor.
— Green leaves exhale oxygen in sun light and carbon dioxide in the dark.
— Photosynthesis is the production of sugar from water and carbon dioxide in chlorophyll
    of the plant leaf.
— Chlorophyll production needs sunlight.

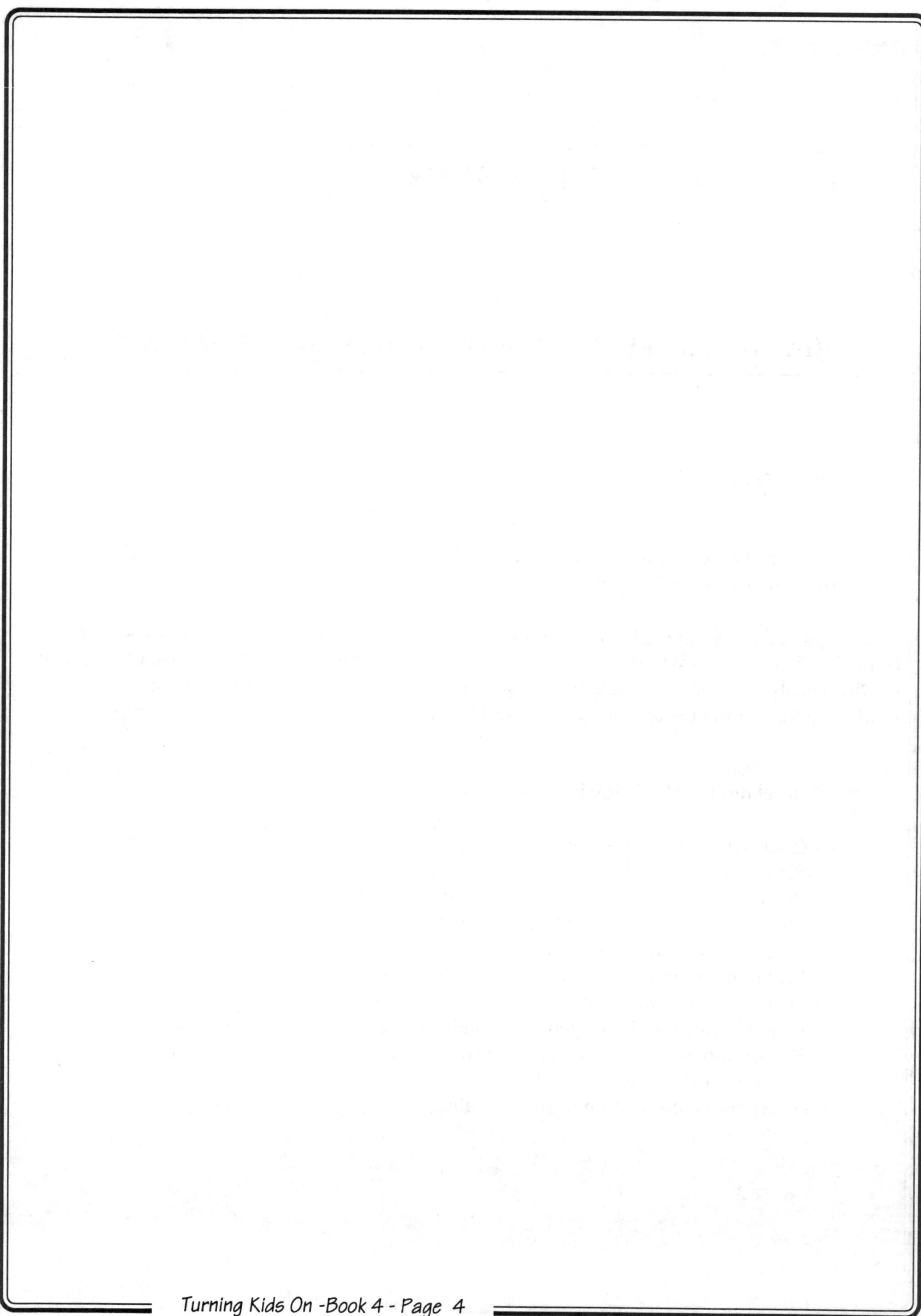

4.1.01.   HOW DO SEEDS GERMINATE ?

What you need: 1. Small seeds (mustard, radish, green beans, etc.).
              2. Blotting paper or several layers of paper towel.   3. A drinking glass or beaker.

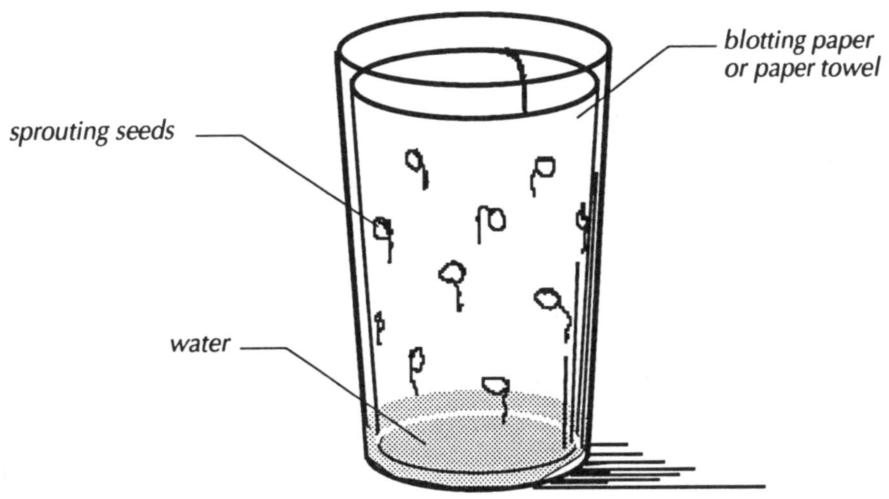

How to do it:
1. Cut a rectangular piece of blotting paper as wide as the glass is tall (if paper towel is used, triple or quadruple the layers).
2. Moisten the paper and place the seeds on it.
3. With the seeds sticking to the moist paper, roll the paper in a cylinder slightly smaller in diameter than the glass (with seeds on the outside) and insert the roll into the glass.
4. Let the paper stand against the glass wall and pour some water into the glass to keep the paper moist.
5. Put the glass in a warm place and cover loosely (make sure that some water stays in the glass at all times).

What to ask:
1. Which seeds sprout first? Which last?
2. In which direction did the rootlets grow?
3. How could the paper stay moist at all times?
4. Why did the glass have to be loosely covered?
5. What are the conditions for germination of seeds?

How to explain it:
    In three to six days, depending on what seeds, the seeds will **sprout** and send their rootlets in a downward direction: this is in the direction of the water. The blotting paper stayed moist, because the water worked itself up through **capillary action** in the paper fibres.
    The glass has to be covered in order to prevent fast evaporation of the water, but it has to be done loosely so that the seeds are not cut off from the atmospheric air.
    The ideal conditions for germination of seeds are: presence of **moisture**, **warm temperatures**, and supply of **air.**

4.1.02. THE BENDING PLANT

<u>What you need</u>: 1. Bean or radish seeds. 2. A small flower pot and soil.
     3. A cardboard box.

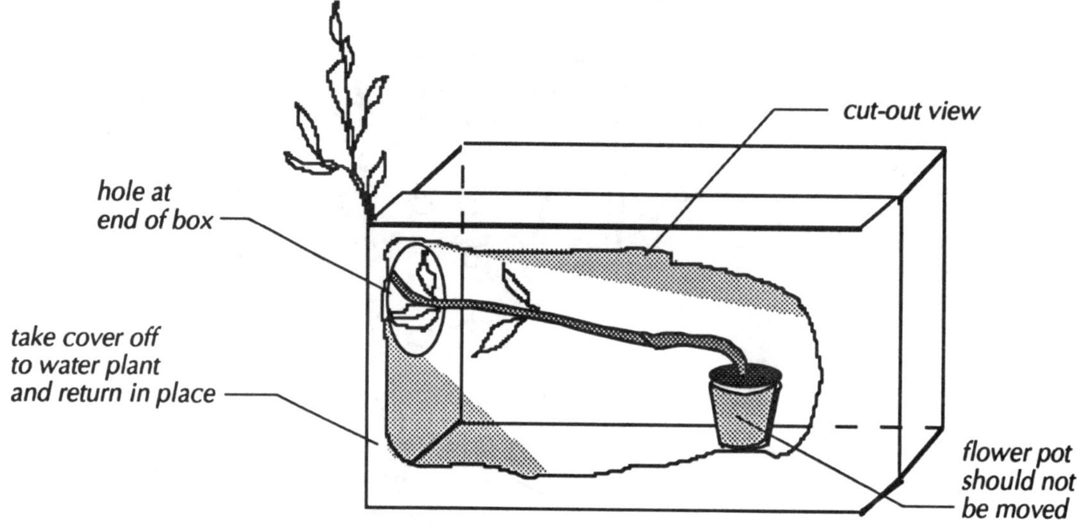

<u>How to do it:</u>

1. Plant one seed on one small flower pot and water daily.
2. After the seed has started to sprout, let it grow until it is about 5 cm high.
3. Now place the pot in the cardboard box in which a hole has been cut out in its end (see sketch).
4. Keep the box close to a window to let light in through the hole.
5. Keep the pot in the same position, water daily for the next few days.
   (replace cover after watering), and observe plant growth!

<u>What to ask:</u>

1. How did the plant grow? In which direction?
2. How would the plant grow, if the pot were turned around (180°) every day?
3. How would the plant grow, if the box were turned around (180°) every day?
4. How would the plant grow, if the box cover were left open?
5. What do plants need in order to grow?

<u>How to explain it:</u>

  Plants grow in the direction from where the light comes, in this case towards the hole in the box, and even through the hole when left for a longer time. This shows that the plant needs light in order for it to grow. By turning toward the light (hole in the box) its leaves will catch more light and thus grow better.

  If the plant were turned half way around every day, it would grow straight up, or keep bending back and forth. If the box were turned, the plant would still bend towards the hole. Without the box cover, the plant would just grow straight up (and slightly bend toward the window).

4.1.03. THE CROOKED ROOT

<u>What you need</u>: 1. A small young plant and potting soil.
2. A flat, square, transparent container or two, 10 x 10 cm pieces of window glass & three 1 x 1 x 10 cm wooden blocks.   3. Cardboard or black paper.

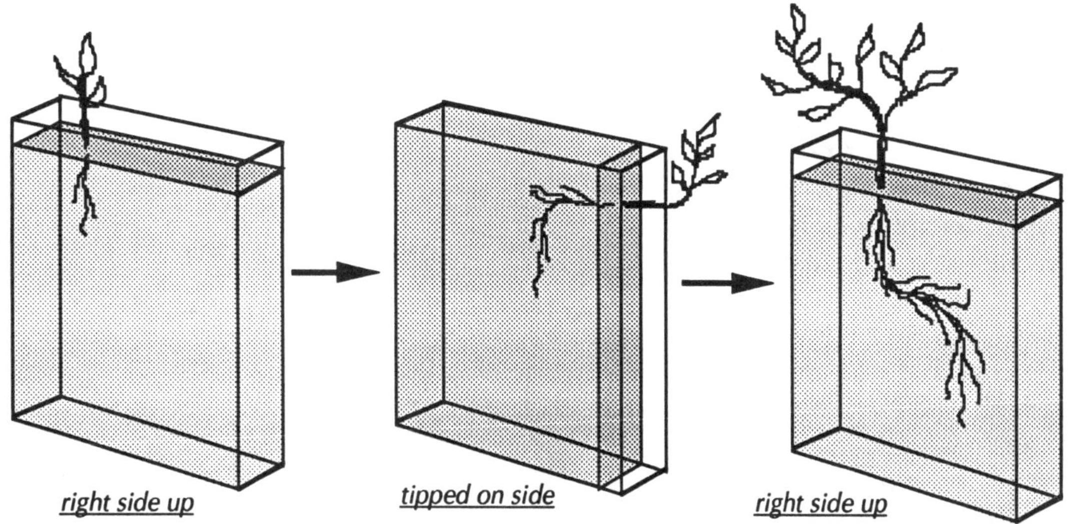

<u>How to do it</u>:

1. If a transparent thin square container (chemical cell) is not available, construct one by taping two pieces of window glass against each other with the three wooden blocks in between them
2. Fill the container with potting soil and place the plant on one side in the soil of the container.
3. Cover the sides of the container with cardboard or black paper and let the plant grow for 3-4 days.
4. Tip the container on its side (cover the open top with cardboard if the soil tends to fall out), and let the plant further grow.
5. After another 3-4 days, turn the container right side up again.
6. Point 4 and 5 can be repeated once more, then remove the covering cardboard or black paper and observe the roots.

<u>What to ask</u>:

1. How did the roots of the plant grow?
2. In what direction do roots grow?
3. Why did the sides of the container have to be covered?
4. On the day just before turning the container right side up, in which direction did the plant grow?

<u>How to explain it</u>:

Plant roots grow downward in a **vertical** direction. They grow towards the center of the earth, opposite to the direction of the plant parts that are above the ground, which grow vertically upwards. Regardless of the position of the container, the roots will grow downward, this is why turning the container on its side makes the roots turn in their direction of growth. The covering of the sides of the container keeps the roots in the dark, simulating the earth or an opaque pot of soil.

4.1.04. | THE WATER-SUCKING ROOTS

What you need: 1. A beaker (250 ml) or drinking glass, a one-hole stopper, a glass tube.
2. A carrot or a cylinder shaped potato, syrup (sugar), candle wax.
3. A coring knife (apple corer), a stand and clamp.

How to do it:

1. With the coring knife, cut a hole in the carrot or potato about three quarters down its length, such that the one-hole stopper will fit in it and close it tightly (see sketch).
2. Insert a 20 cm long glass tube in the one-hole stopper.
3. Fill the hole in the carrot or potato with syrup or a concentrated solution of sugar in water brimful.
4. Push the stopper with the glass tube in the hole (liquid level should rise in the tube) and seal any openings between the stopper and the carrot or potato with candle wax (light a candle and let the melted wax drop on the places that you want sealed).
5. Mark the liquid level in the glass tube with a piece of masking tape , a grease pencil, or a rubber band.
6. Clamp the carrot or potato and immerse it in water. Observe the water level in the glass tube at the end of the period.

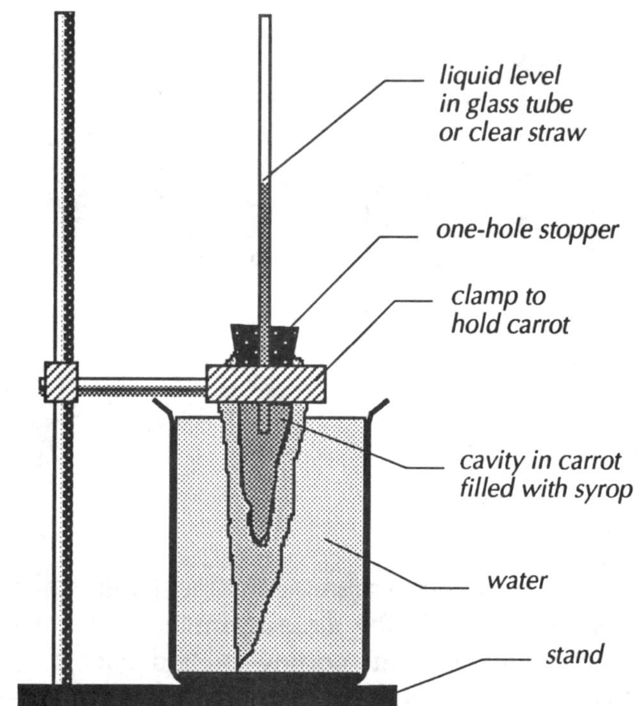

What to ask:

1. What made the water level in the glass tube rise?
2. Would this water level also rise if the tube were filled with plain water? With salt water?
3. Why did the stopper have to be sealed with wax?
4. What would happen if the carrot and tube were filled with plain water and the beaker with sugar solution?

How to explain it:

The skin, tissue, and fibres of the carrot or potato act like a **semi-permeable membrane** letting only the small water molecules through, but not the large sugar molecules. This makes the water move from the beaker into the carrot and up the tube. If the concentration of sugar is higher in the beaker compared to that inside the carrot, the water will move out of the carrot and thus the water level in the tube will go down. This action and migration of water molecules through a semi-permeable membrane is called: **osmosis** and the resulting pressure: **osmotic pressure.**

4.1.05. | MAKE AN EGG OSMOMETER

What you need: 1. A fresh chicken egg, soda straw (transparent) glass tubing.
  2. Dilute HCl or strong vinegar, household cement or sealing wax.
  3. A saucer, a drinking glass, a thin wire.

How to do it:

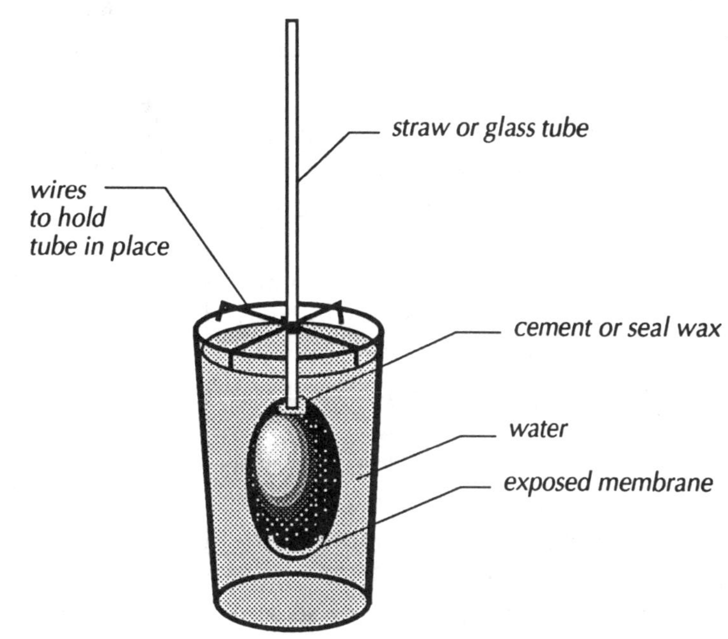

1. Put some dilute hydro-chloric acid or strong vinegar on the saucer and hold the large end of the egg in the liquid until the shell has been eaten away, leaving the thin membrane exposed.
2. After an area of about 4 cm² of the thin membrane has been exposed, rinse the acid off with fresh water.
3. At the other end of the egg, make a hole the size of the straw or glass tube (take care that the egg does not crack).
4. Insert the straw or glass tube through the hole into the interior of the egg and seal any excess opening between the egg and the tube off with cement or wax (candle wax will do).
5. Place the egg osmometer in a glass of water and hold the tube in place (vertically up) by wrapping the thin wire around it and bending the wire over the rim of the glass (see sketch).
6. Let stand for a few hours and observe liquid level in the tube.

What to ask:
1. What do you think the egg shell consist of?
2. Why does the seal between the egg and the tube have to be absolutely tight? What would happen if it were not?
3. Why did the liquid level in the tube rise?

How to explain it:
  The thin membrane of the egg acts like a **semi-permeable membrane**, which allows small size molecules of water to pass through it, but not the large protein molecules in the egg. The water thus **migrates** from the outside of the egg to the inside, and in so doing pushes the liquid level up the tube. This liquid level difference, between that in the tube and that of the water in the glass, is called **osmotic pressure.**

4.1.06.   THE SWOLLEN EGG

What you need: 1. A fresh chicken egg.   2. Dilute HCl or strong vinegar, a cup or beaker.
3. An old fashioned milk bottle (or wine carafe or any bottle of which the mouth diameter is a bit smaller than the egg's).

How to do it:

1. Place the egg in the cup and add dilute HCl (hydrochloric acid) or strong vinegar to it until it is completely immersed. Hold the egg under the surface of the liquid or keep rotating the egg. Do this for 10 minutes.
2. Check whether all the egg shell has dissolved by touching the egg and very carefully pressing on it. When only the membrane is left, it should feel soft and flexible.
3. Fill the milk bottle about 3/4 full with hot water (almost boiling) and place the egg immediately on the mouth of the bottle and let stand.
4. Observe! Leave for an hour or so and observe again!

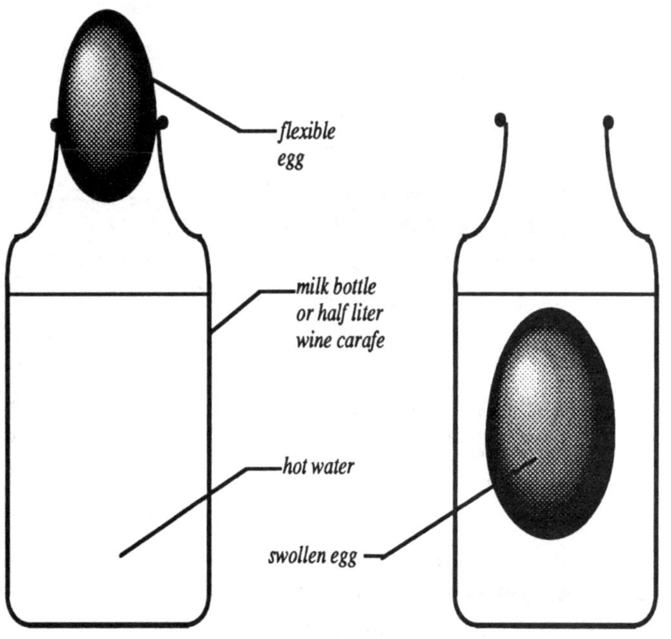

What questions to ask:

1. why did the egg become soft and flexible after placing it in acid?
2. What did the hot water do to the air in the milk bottle?
3. What did the flexible egg do after it was placed on the bottle?
4. What did the egg do after an hour wait?
5. What made the egg grow larger in the bottle?
6. How can we get the egg out of the bottle without breaking it?
7. How can we reverse the process of osmosis?

How to explain it:

The shell of an egg consists mainly of calcium carbonate. By placing the egg in dilute hydrochloric acid or strong vinegar, the calcium carbonate reacts with the acid to form a soluble calcium salt and carbondioxide gas. This leaves only the membrane around the egg, which makes the egg feel soft and flexible.

The steam of the hot water drives the air partly out of the bottle, thus by placing the flexible egg on the bottle, the egg is slowly sucked into the bottle while the water is cooling off.

After the egg is totally plunged in the water, **osmosis** is taking place. Water molecules are migrating through the **semi-permeable membrane** of the egg from the outside into the egg and thus making the egg swell. In order to get the egg back out, we have to shrink the egg, which can be done by pouring all the water out and letting the egg dry out and shrink. Then by heating the bottle while the egg sits in the neck, it will slowly be pushed out by the expanding air.

4.1.07. | HOW DOES GRAVITY AFFECT GROWTH ?

What you need: 1. Three small plants (identical in size and kind).
                2. A plant hanger and hook, cardboard.

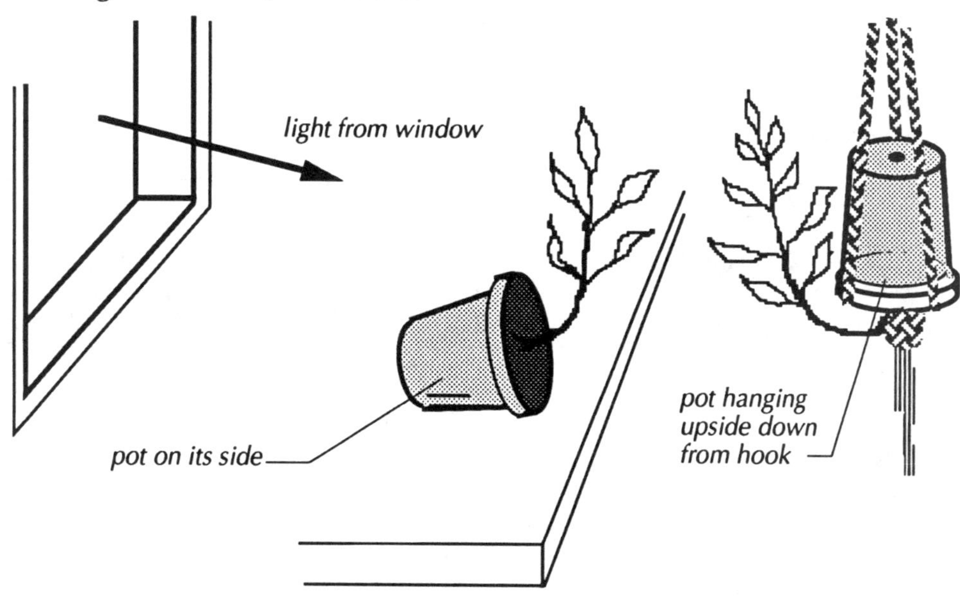

How to do it:
1. Turn one flower pot with the small plant in it on its side and face the plant away from the window.
2. Place another flower pot with the small plant upside down on a hanger (tape cardboard securely over soil to prevent it from falling out), and hang this from a hook close to the window.
3. Place the third flower pot with plant normally in front of the window.
4. Observe plant growth in all three pots and compare.

What to ask:
1. How did the plant on its side grow?
2. How did the upside down plant grow?
3. In what direction did all three plants grow?
4. Why did the plant on its side have to be faced away from the window?
5. How did the light from the window affect the growth?
6. Can we be absolutely sure that the light was not the cause of the plants to bend upwards?
7. How would we set up an experiment to investigate the influence of only gravity on the growth of a plant?

How to explain it

Plants in general grow in the opposite direction of the growth of the roots, which is vertically upwards. They grow **in the opposite direction of the gravity force.** In nature, we can especially notice this when trees are growing on a steep hill or on the side of a mountain or cliff.

An experiment investigating the influence of gravity **only** would need to have all variables controlled, including the light variable, which has to come from all sides toward the plant. An open-ended question would be: How would a plant grow without gravity?

4.1.08. MAKE SOOT PRINTS OF LEAVES

What you need: 1. Two empty jars with tight lids.   2. A candle, grease or Vaseline.
    3. Old newspaper , white blank paper, different leaves.

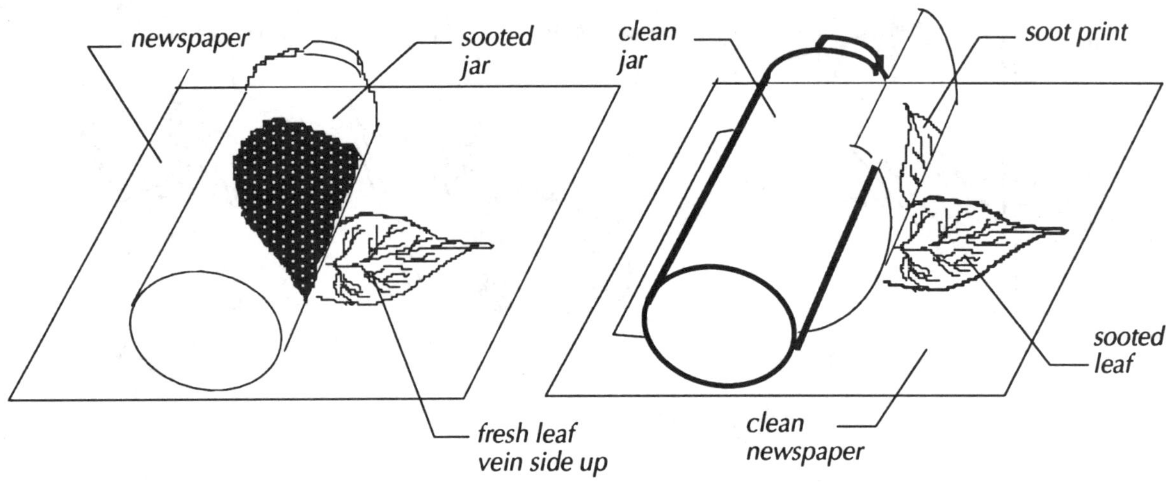

How to do it:
1. Cover the side of a smooth jar with a thin layer of grease or vaseline. Fill the jar with cold water and screw the lid on tightly.
2. Cover the greased part of the jar with soot by holding it over a candle flame until it is evenly black.
3. Place a leaf with the vein side up on a newspaper and roll the soot-covered part of the jar over the leaf (see Sketch A).
4. Take the leaf and place it on a clean newspaper. Cover the leaf with a sheet of white paper and roll a clean jar over the paper and leaf. (see Sketch B).

What to ask:
1. What is the purpose of the grease or vaseline layer?
2. Why did we need cold water in the first jar?
3. Why was the leaf placed on the newspaper vein side up?
4. Which part of the leaf was covered with the black soot?
5. What type of leaves would give the best soot prints?

How to explain it:
The jar with the grease and soot layer functioned as an **ink roller** to cover the veins of the leaf with a thin layer of soot. The leaf had to be placed vein side up, because the veins are bulging up (protruding) compared to the other parts of the leaf, and this makes it possible to color only the veins. When the white paper was placed on the sooted leaf, the veins only would give a black print on the paper. Leaves with strong bulging veins make the best soot prints.

4.1.09. | DO LEAVES GIVE OFF WATER ?

What you need: 1. A large wide leaf or small plant.
2. Four large drinking glasses (of the same size).　　3. Two paper cards.

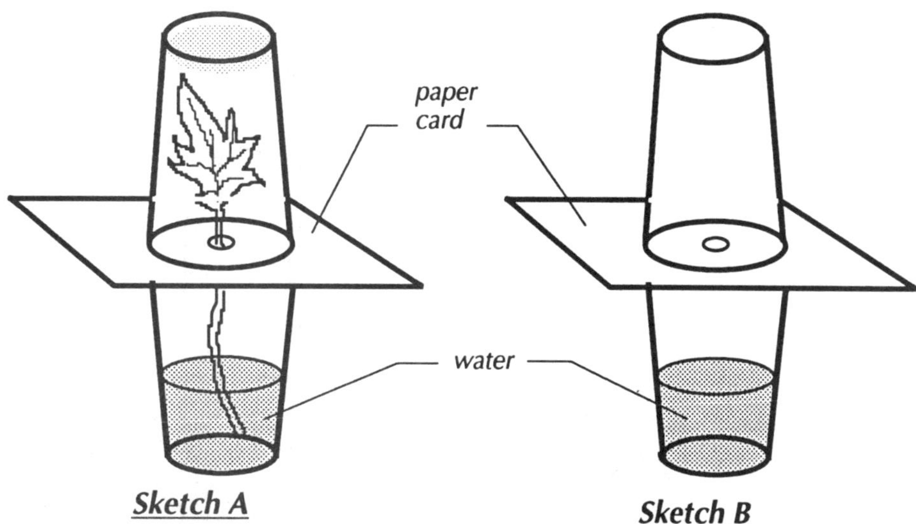

How to do it:
1. Fill two of the glasses halfway with water.
2. Cut a small hole in each of the two paper cards, such that the stem of the leaf or small plant would fit through it.
3. Cover the glasses with the paper cards, insert through one card the leaf stem, and cover with the inverted glasses (see sketch).
4. Observe and compare the inside of the upper glasses.

What to ask:
1. In which glass did water droplets form?
2. Why did we need the set of glasses in Sketch B?
3. Where did the water droplets come from?
4. What kind of leaves would give off most water?
5. What variables would influence the amount of water transpired?
6. Would a whole plant covered in the same manner give off water too?

How to explain it:
　　The water **transpired** from the leaf is in the form of **water vapor**. This latter condenses on the cold surface of the upper glass in Sketch A and thus water droplets are formed. In order to show that these droplets are indeed coming from the leaf rather than the water in the lower glass, the set-up in A is being compared with one without a leaf in it (see Sketch B).
　　A small plant covered in the same way with a card around the stem in the pot and an inverted glass over the whole plant, would give the same results.
　　Variables that would influence the amount of water transpired are: type of leaf, surface area of the leaf, age of leaf, etc.

4.1.10. | WHAT DO GREEN LEAVES BREATHE OUT ? (I)

What you need: 1. Green weed & wood split.
   2. A large beaker, a funnel, a test tube.   3. A stand and clamp.

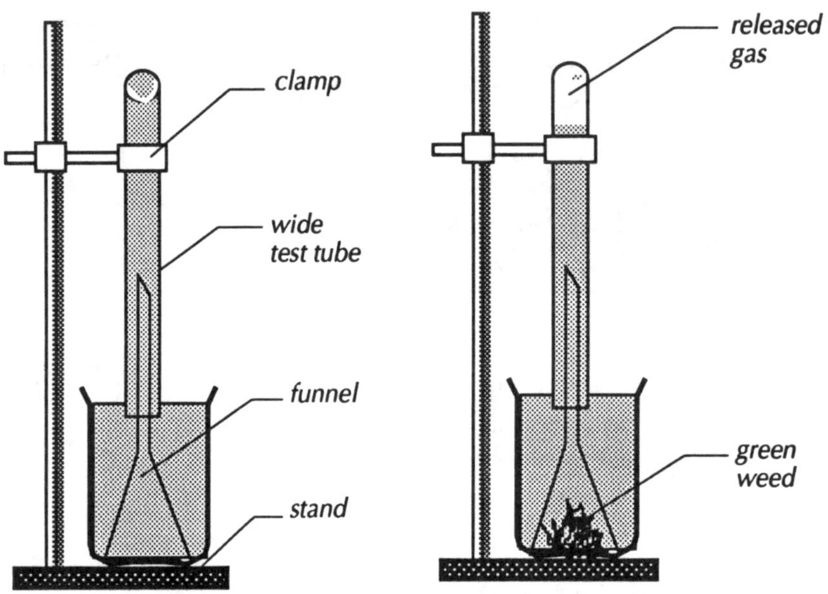

How to do it:
1. Fill the beaker with water, immerse the funnel and the test tube in the water, and set the apparatus up as in Sketch A.
2. Raise the funnel and place some green weed under it.
3. Leave the apparatus in strong sunlight or under a spotlight and observe the bubbles given off by the leaves.
4. After collecting almost a full test tube of gas, test it with a glowing wood splint.

What to ask:
1. What gas is collected in the test tube?
2. What did the glowing wood splint do when lowered in the test tube?
3. What made the water in the test tube stand so much higher than the water level in the beaker? (Sketch A).

How to explain it:
The green in the leaves, which is **chlorophyll**, produces **sugar** and **cellulose** and **starch** in the plant. During this process of sugar production **carbon dioxide** and **water**, **oxygen** is released. This only occurs during daytime when the sunlight is shining on it. The purpose of the funnel is to bring all the bubbles released by the weed together under the test tube. As the glowing splint flares up into the bright flame in the gas, it indicates that the gas is **oxygen**. The fact that plants give off oxygen during the daytime makes having them in the living roam a good thing. The air is enriched with oxygen and it is therefore healthy to have plants in the room.

4.1.11. | WHAT DO GREEN LEAVES BREATHE OUT ? (II)

What you need: 1. A wide-mouthed gallon jar & lid.   2. A small potted plant.
3. Limewater (calcium hydroxide).

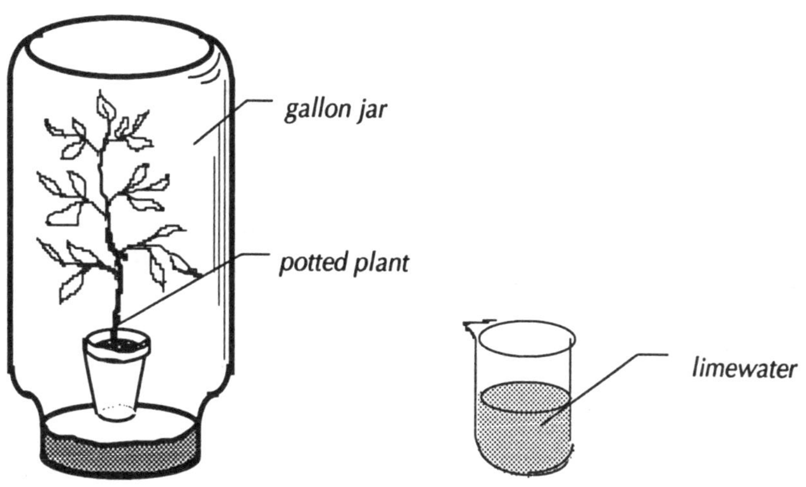

How to do it:
1. Place the potted plant on the inverted lid of the jar and invert the large jar over the whole plant, and turn the jar into the lid (see Sketch).
2. Leave the jar and plant in a completely dark place for several hours.
3. Turn the lid loose and remove the plant carefully.
4. Turn the jar right side up and pour about 100 ml of limewater in it, screw the lid on, and shake.
5. Observe the limewater (turning milky).

What to ask:
1. What gas was released by the leaves in the dark?
2. What reaction did the gas give with the limewater?
3. How would this gas react with a glowing wooden splint?
4. With many plants in a room, how would the composition of the air change from daytime to nighttime?
5. Why is it better to take plants out of the bedroom at night?

How to explain it:
Plants give off **carbon dioxide** when placed in the dark. This is indicated by the reaction with the limewater (calcium hydroxide): $CO_2 + Ca(OH)_2 = CaCO_3 + H_2O$, forming calcium carbonate which is insoluble in water. This is the cause for the limewater to turn milky.

When this gas, produced in the dark, is tested with the glowing splint, it would extinguish the glow completely, as carbon dioxide does not support the burning process. With many plants in a room, the air would be rich in oxygen content during the day and high in carbon dioxide content at night. This is why it is better to take plants out of the bedroom at nighttime.

4.1.12. | HOW IS THE GREEN IN THE LEAVES PRODUCED ?

What you need: 1. A plant with large wide leaves.
2. Carbon paper or black construction paper.   3. Paper clips or masking tape.

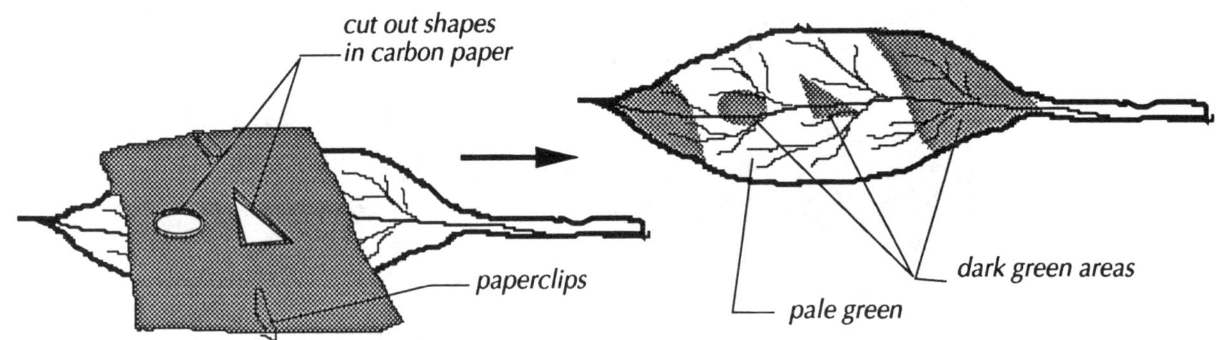

How to do it:

1. Cut out several patterns (circle, square, triangle) in several pieces of the carbon paper.
2. Cover three or more leaves as much as possible with the cut out carbon paper by attaching it to the leaves with the paper clip or masking tape.
3. Cover some leaves halfway with carbon paper close to the stem (or any other pattern of covering) and leave it attached for 2 or 3 days. (do this with living leaves)
4. After leaving the black paper against the leaves for several days, remove the attached paper and observe the leaves.

What to ask:

1. How did the covered areas of the leaves compare to the uncovered ones?
2. Do plants need sunshine to produce the green color?
3. What is the green color in the plant leaves called?
4. What is the process of production of the green color called?
5. What is the function of the chlorophyll in plant leaves?

How to explain it:

The covered areas of the leaves will become much paler. The longer it stays covered, the paler the color, because no sunshine is penetrating the green pigment that enables every plant that possesses it to combine water and carbon dioxide from the air to form sugar. This process in which sunshine is an essential ingredient is called **photosynthesis**. It is the sugar in the plants which gives animals and man the energy when it is consumed by them.

The **chlorophyll** also produces cellulose, a much larger molecule than sugar, which is the basic building material in plants. Thus, without sunshine the leaves do not produce chlorophyll, no cellulose, and therefore plants do not grow.

4.1.13.   CAN AIR ENTER THROUGH A LEAF ?

What you need: 1. A leaf with a long stem.   2. A small Erlenmeyer flask.
3. A 2-hole stopper that fits in the flask.   4. A bent glass tube.   5. A candle & matches.

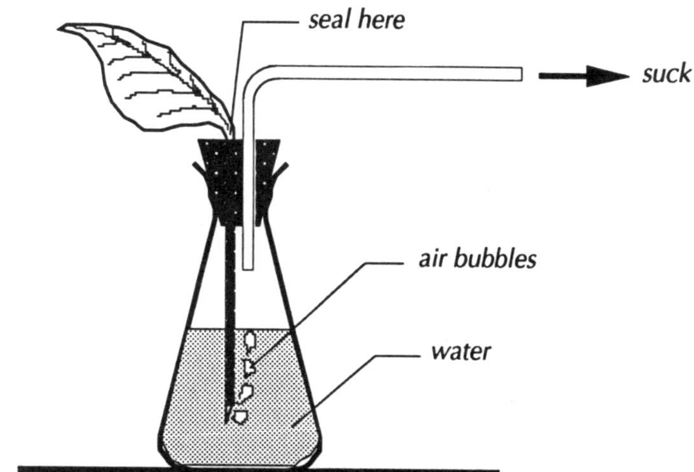

How to do it:
1. Stick the leaf stem through one of the holes in the 2-hole stopper and seal it with dripping wax from a lit candle.
2. Insert the bent glass tube in the other hole of the stopper.
3. Fill the Erlenmeyer flask with water to such a level that only the leaf stem immerses in it and not the glass tube.
4. Place the stopper tightly into the flask and suck through the side tube.
5. Observe air bubbles issuing from the end of the stalk.

What to ask:
1. Why does the stem have to be sealed in the stopper?
2. What would happen if the glass tube were also immersed in the water?
3. Are the leaf and stem actually that porous that air can go through them?
4. What is the actual structure of the leaves?

How to explain it:
    The sucking through the side tube **lowered the pressure** inside the flask, causing the atmospheric air to seep through the leaf and the stalk resulting in the bubbles issuing from the end of the stalk. When looking through a microscope and examining the underside of leaves, we can see breathing pores or **stomata** with two little guard cells on each side of the stomata (see sketch on right).

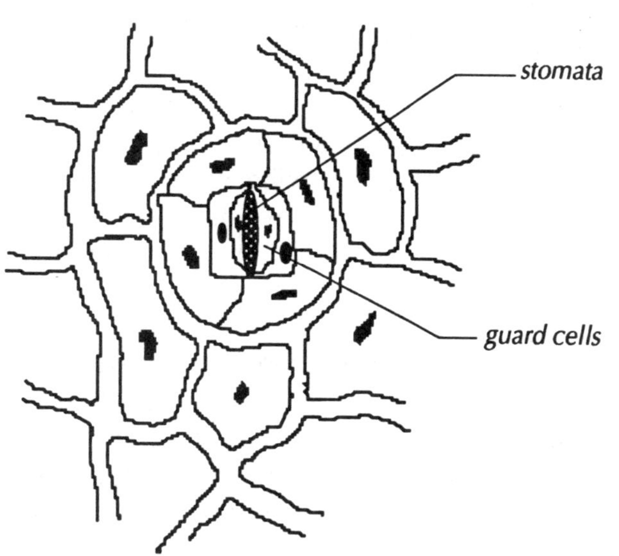

4.1.14.   GROW SWEET POTATO AND CARROT LEAVES

What you need: 1. A sweet potato and/or carrot.   2. A drinking glass or beaker.
   3. A shallow disk (saucer).   4. Three toothpicks or thin nails.   5. Some pebbles or gravel.

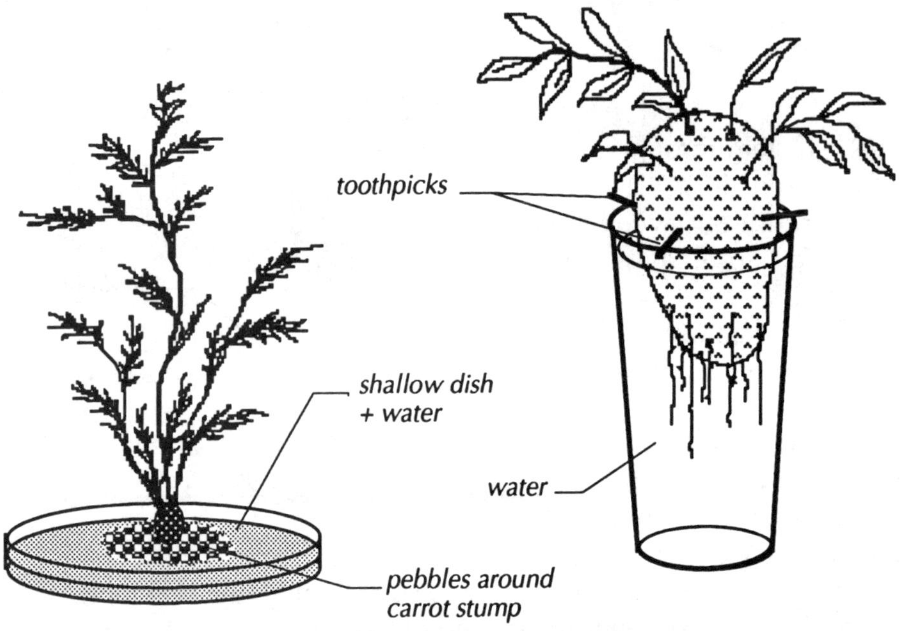

How to do it:
1. Hold the sweet potato with the eyes or buds on the top side.
2. Immerse the potato about one-third in the water contained in the drinking glass.
3. Support the potato by sticking three toothpicks or nails into its side and resting it on the glass rim.
4. Remove the old leaves from the top of the carrot and cut off all of the lower part except about 5 cm.
5. Place this stump in water in the shallow dish and support with pebbles or gravel around it.
6. Put sweet potato and carrot in a warm and sunny place and observe foliage growth.

What to ask:
1. Where did the potato or carrot plant get its food from?
2. How does the potato leaf differ from that of the carrot?
3. Would these plants develop new potatoes or carrots?
4. What other plants could be grown this way?

How to explain it:
The potato, like the carrot, beet or turnip, contains much **stored food**. When placed in water, it is enough to produce thick foliage. When this food supply is depleted, it needs to be placed in soil and receive other nutrients in order to produce new potatoes and carrots. When a pineapple is cut about 5 cm below the base of the leaves and placed in water, the leaves will continue to grow for quite some time drawing nutrients from its stored food in the pineapple itself.

4.1.15. | MAKE A RED-BLUE CARNATION

What you need: 1. A white carnation (with a long stem).
    2. Red food coloring & blue ink.   3. Two small beakers or cups.

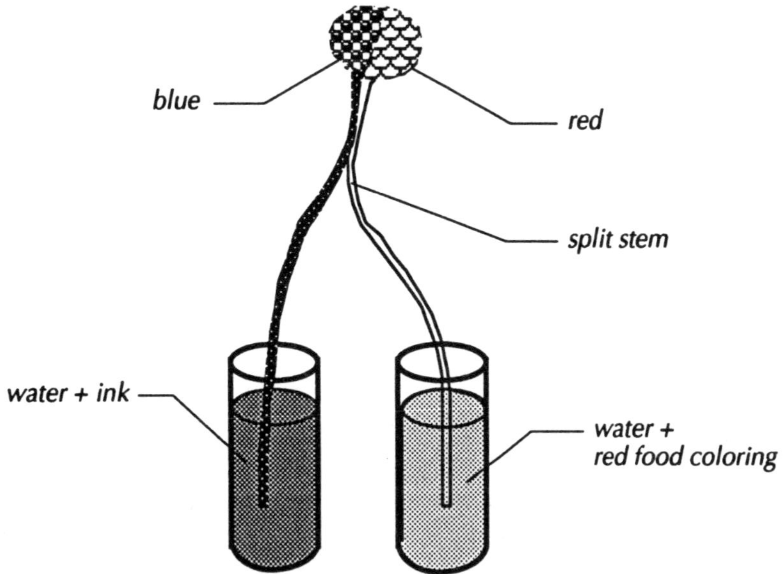

How to do it:
1. Take the carnation and cut the stem to leave about 25 cm on the flower.
2. Split the stem in half, starting a cut with a knife and further splitting it along the fibres (without breaking them!)
3. Fill the two small beakers with water, color one red with the food coloring and the other blue with the ink (the color should be very intense).
4. Place each half of the stem in each beaker (support it by taping) and observe the flower.

What to ask:
1. What was the purpose of splitting the stem in half?
2. Did the whole white carnation get colored?
3. Is it possible to color only part of the flower and leave another part white?
4. How would the stem have to be split to obtain three colors in the flower, say red, blue and white?
5. What force is pulling the colored solution up the stem?
6. Would leaves be able to be colored the same way?

How to explain it:
The colored water is drawn up the stem of the carnation by **osmosis** and **capillary action**. The water molecules **diffuse** through the fibre membranes from a less to a larger concentration of plant sap (**osmosis**). The fibres are so tiny that the adhesive force of the water molecules to the fibre walls becomes very great. This capillary force in combination with the **osmotic pressure** sucks the water up the flower.

When the stem is split three ways, it is very likely that the flower will be three way colored. This shows that the fibres must somehow run all the way from the stem to the petals of the flowers.

4.1.16. THE TOOTHPICK STAR

What you need: 1. Five toothpicks (flat kind), 1 box for the whole class of 30.
  2. A cup of water.

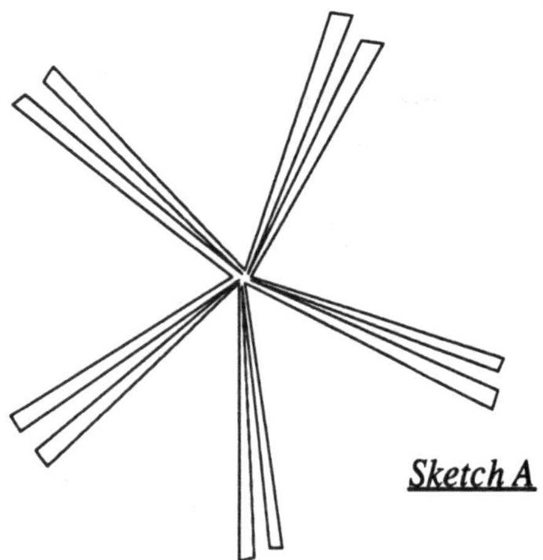

Sketch A

Sketch B

How to do it:

1. Break five toothpicks in half but leave the parts together, and leave them in a V-shape.
2. Place the five V-shaped toothpicks on a smooth surface with their points as close as possible together (see Sketch A).
3. Dip your finger tip in the cup of water and let one drop fall in the center of the toothpick configuration.
4. Observe carefully what is happening to the toothpicks!

What questions to ask:

1. What happened to the toothpicks?
2. What did the water do with the toothpicks?
3. What material do the toothpicks consist of?
4. What will happen to picked flowers when they are not placed in a vase in water? Why do they need water?
5. How does water reach the top leaves in a tall tree?
6. Ccmpare a dry broken toothpick with one that was wetted by holding one of the legs of the V-shape and pushing the other leg down: Which one stays closed/down?

How to explain it:

By breaking the toothpick in half, not all the vessels/fibers are cut off. The water is **absorbed** by the wood fibers by **capillary action,** which consists mostly of adhesive forces. Also **diffusion** through the fiber walls is taking place: water is thus filling the unbroken vessels or fibers. This makes the toothpick tend to straighten up and make it more springy than the dry one.

This is the reason why picked flowers become limp when they are not placed in water. Capillary action and **osmotic pressure** are the forces that bring the water from the roots of a tree all the way up to the top leaves.

When repeating the activity, dry toothpicks have to be used. The wet ones will not form and stay in a narrow V-shape. They become more springy than the dry toothpicks and form a much wider V-shape.

4.1.17. | HOW DO MOLDS REPRODUCE ?

What you need: 1. Eight slices of bread.   2. Saran Wrap or cellophane.
3. A magnifying glass.   4. A shallow dish (saucer or petri dish).

How to do it:

1. Take two slices of bread. Wrap one slice in the Saran Wrap and seal off from the air and leave the other one open to the air.
2. Take two other slices: place one in a dry spot and place the other over a shallow dish filled with water.
3. Expose two other slices: one to a strong light (flood light) and keep the other in a dark place.
4. Expose two more slices: keep one in a warm dark place and the other in a cold dark place.
5. Observe the surface of the bread slices daily through the magnifying glass and record the changes. Make a note of the mold colonies, the number of them and their growth rate.

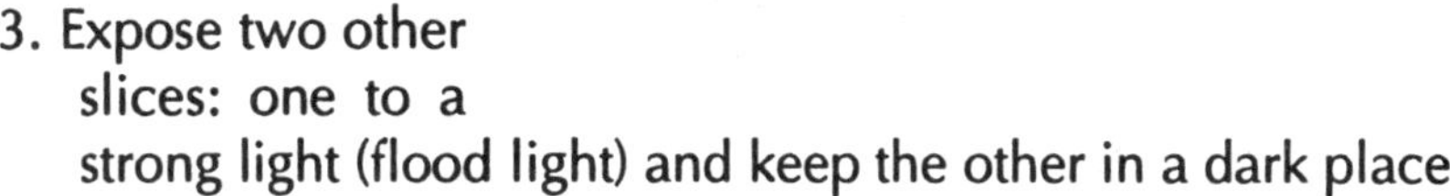

What to ask:

1. What variable was manipulated in step 1 of the procedure?
2. What variable did I choose to change in Steps 2, 3, and 4?
3. What was the responding variable in all four cases?
4. In which case did the mold develop best?
5. How should bread be stored to keep it mold free?
6. What other types of food will develop molds if kept over a longer period of time?

How to explain it:

Mold grows best in moist, dark and warm places. Dry, light and cold conditions will discourage the growth of molds. Under step 1 of the procedure the presence or absence of air was the **manipulated variable**, under step 2, moisture, step 3, the presence of light, and step 4, the temperature.

Other foods that will develop molds are: cheese, jam, leftover food or meats, etc.

4.1.18. | PICK UP A CARAFE WITH A STRAW

What you need: 1. A wine carafe (or other long necked bottle).   2. A wheat straw.

How to do it:

1. Fill the carafe half full with water.
2. Bend the straw such that the short bent part will fit in the wider part of the bottle (see sketch).
3. Lower the straw in the bottle and allow the bent part of the straw to open up.
4. Now pull the carafe slowly upwards by holding the protruding straw.

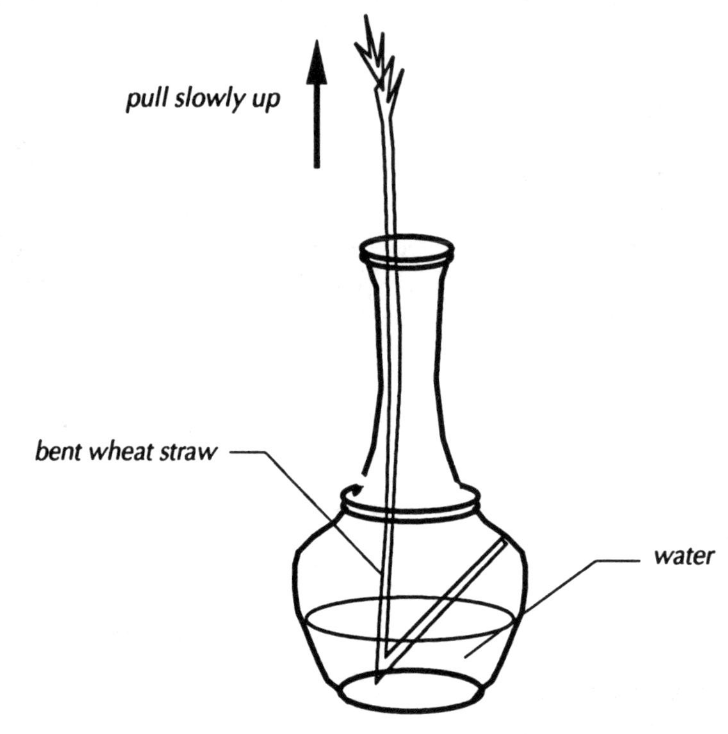

What to ask:
1. Why did the straw not break?
2. What force was exerted on the straw?
3. What will happen to the straw if we wait a while before pulling the carafe up?
4. What will water do when plant material is soaked in it?
5. Where or in what direction lies the greatest strength of a wheat stalk?
6. What implications can we see in living plants?

How to explain it:
It is especially critical where the fold in the straw is made. It should be folded at a spot such that the short part will just fit in the wider part of the carafe. By pulling the straw up, the bottom part of the straw should press on the side, and the top part should hook on the rather horizontal part of the bottle. The stress in the straw would be mostly lengthwise and as the tissue of the wheat stalk is made up of **tubular capillaries** and vessels, it can withstand a rather high stress. Over time, the water will penetrate into the straw tissue and weaken it. This will be more so the case when a dry straw is used rather than a fresh one.

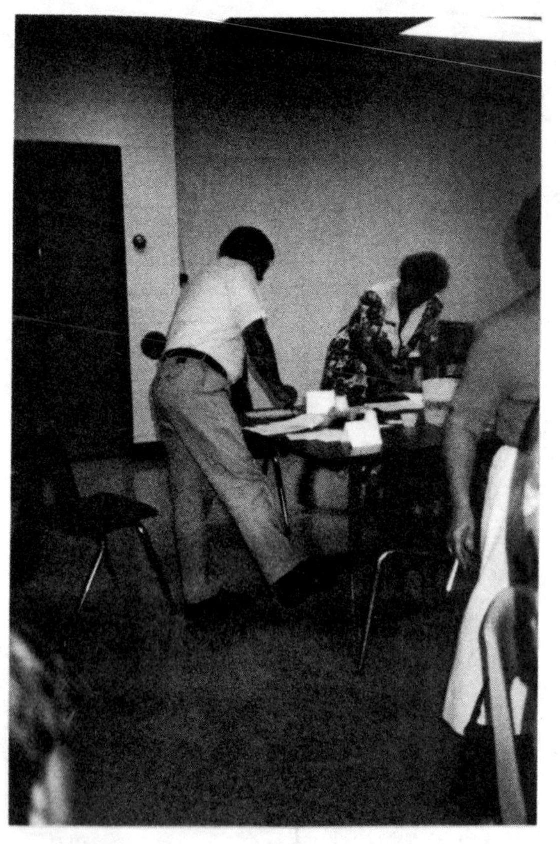# LEARN SCIENCE

BY DOING !

THE UNCOTROLLABLE FOOT

Showing that the left hemi-sphere of our brain controls all the right-hand muscles of our body, and vice-versa.

SEE A HOLE IN YOUR HAND

Showing that the images of both eyes are combined by the human brain.

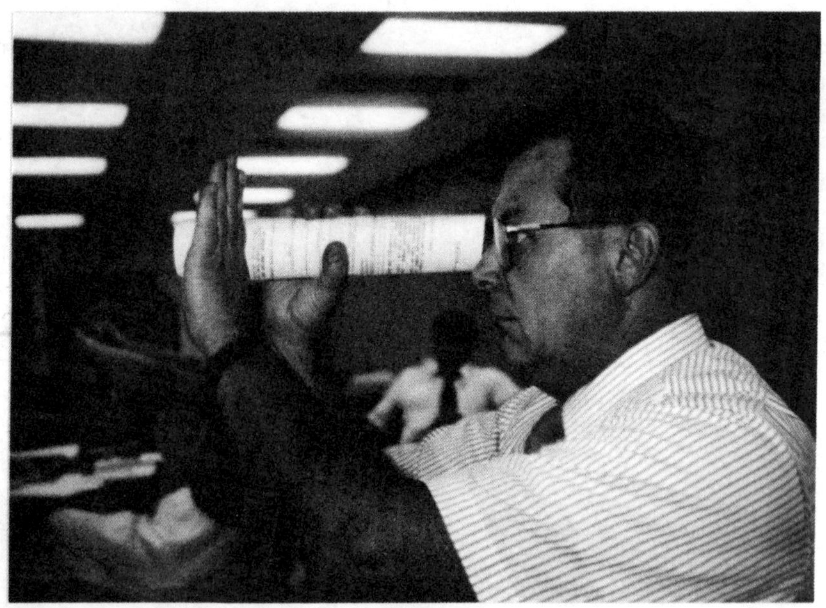

CHAPTER 2

WHAT CAN WE LEARN ABOUT THE HUMAN BODY ?

OBJECTIVES

After dealing with and studying the concepts and sub-concepts of this chapter, the students should be able to:

a. Recognize the correct explanation of an observed event based on each of the sub-concepts;
b. Explain in their own words which of the sub-concepts is determining the course of an event;
c. Distinguish true from false statements concerning each one of the sub-concepts;
d. Identify the correct explanation of an event in daily life applying one of the sub-concepts;

all in relation to the following sub-concepts:

---All objects seen by the human eye cast an upside down image on the retina, which is interpreted by our brain as right side up.
— People need two eyes to see objects in three dimensions.
— The majority of people are right-sighted (possible correlation with right-handedness).
— The cones in the human eye pair the following colors: red-green, yellow-blue, and white-black.
— Illusion is an interpretation of the brain, which deviates from the true perception by the eyes.
---The number of nerve endings is larger at the finger tips compared to other parts of the human body.
---It takes time for an impulse to travel from the receiver (eyes, ears, or other hand) to the arm muscle.
— The human brain is composed of two hemispheres: the left half controls the right-hand side of the body, while the right half controls the left side of the body.
— The number of heartbeats increases with intensity of exercise.
— Exhaled air contains more carbon dioxide than inhaled air.
— Enzymes in saliva break down starch into sugar.
— Food is pushed down the esophagus by peristaltic action.

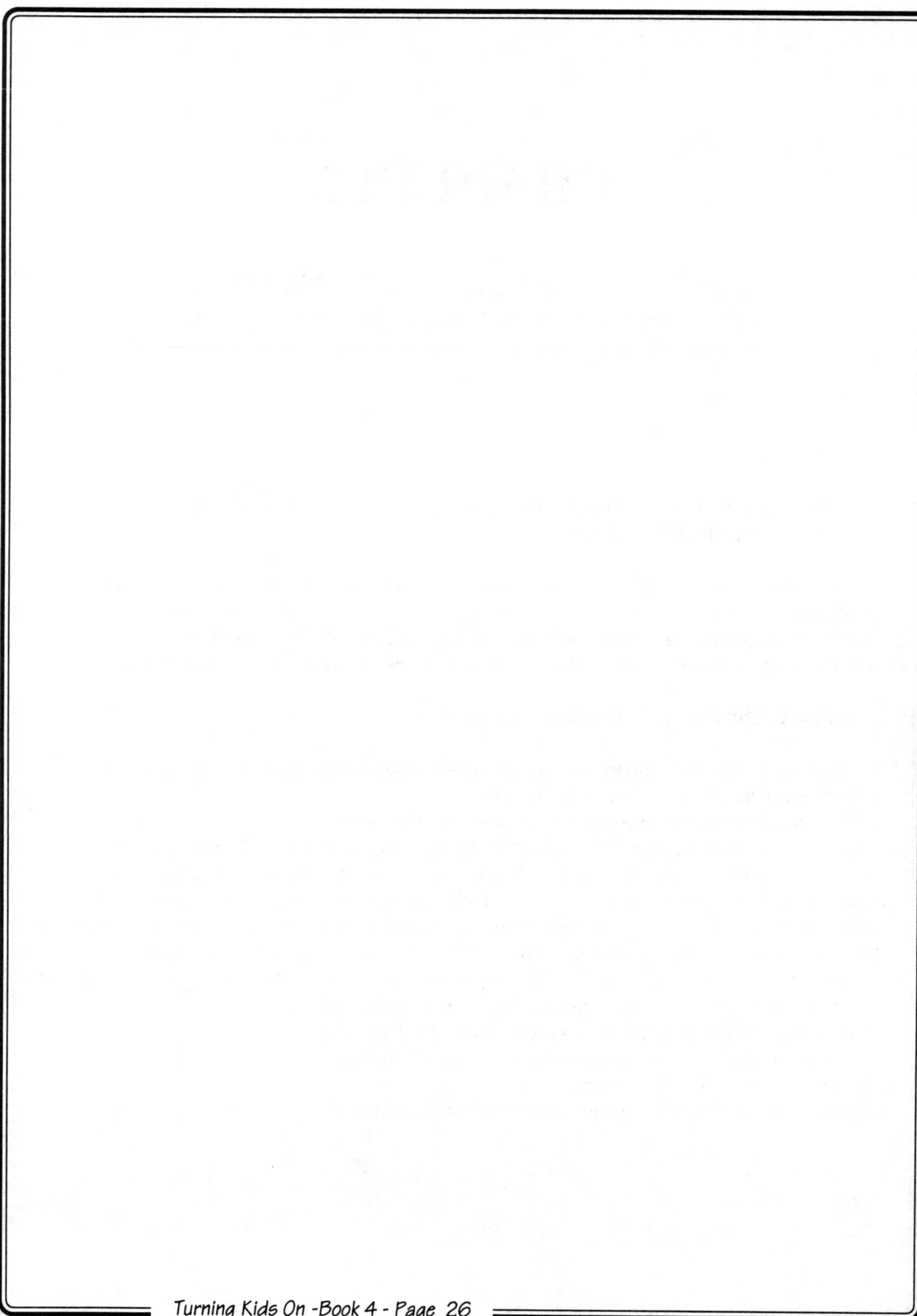

4.2.01. | THE REVERSED IMAGE

What you need: 1. A paper card (3x5"/ 8x12 cm).   2. A regular sewing straight pin.

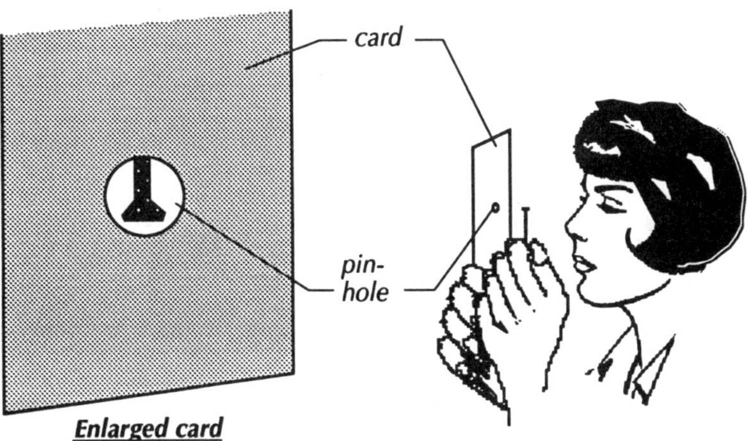

How to do it:
1. Pierce a hole with the pin in the center of the card.
2. Hold the card about 10 cm in front of one eye and hold the pin right side up between the card and the eye (view the pin against a window or light): an upside down image of the pin will be observed.
3. Now hold the hole in the card about 3 cm from the eye and hold your eyelid almost closed: upside down images of the eyelashes will be seen!

What to ask:
1. Why do we see an upside down image of the pin?
2. Why did the image of the pin disappear from in front of the card?
3. What did the eye actually focus on?
4. What was the purpose of the pinhole?
5. Why do we have to view the pin against a window or a light?

How to explain it:
All objects will cast an upside down image onto the **retina** when the human eye is focussed on them, which the brain interprets as right side up (see Sketch A).

The small hole in the card serves as a slit to admit the light, forming a small bundle of light that enters the eye (see Sketch B). This small light beam casts a shadow of the pin upside down onto the retina (actually right side up, but interpreted by the brain as upside down).

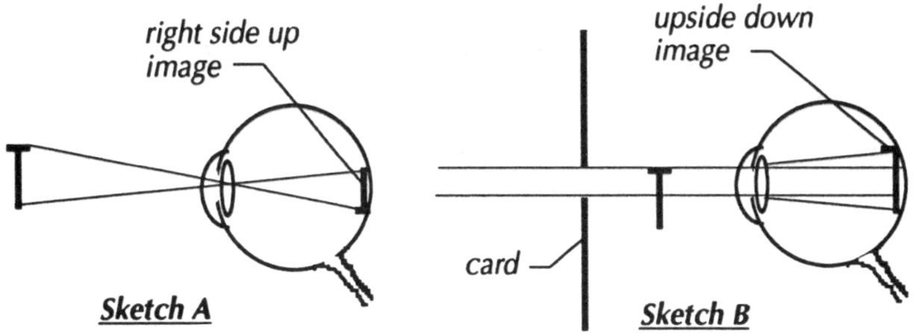

4.2.02.   ARE WE PARTIALLY BLIND ?

<u>What you need:</u> 1. A blank sheet of paper.   2. A pencil or pen.

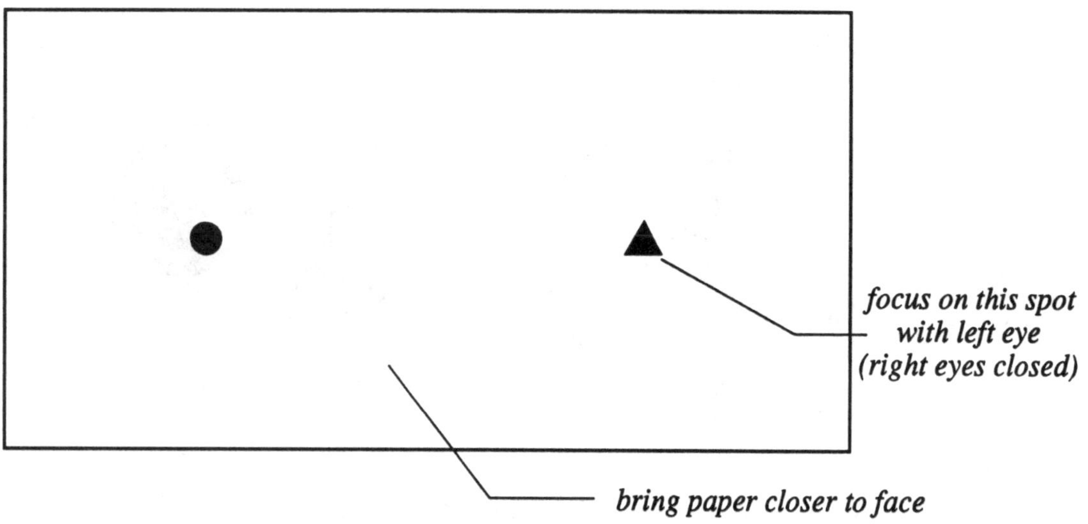

<u>How to do it:</u>
1. Draw a small circle (about .5 cm diameter) and a small triangle (about the same size), about 15 cm apart on the white blank sheet, and completely darken them in.
2. Hold the paper about 30 cm in front of your eyes and shorten this distance very gradually (bring the paper closer to you slowly). While doing this look with your left eye (close right eye) and focus on the triangle or look with your right eye (left eye closed) and focus on the circle.
3. If you do not notice anything happening with the other dot, you may want to hold the paper somewhat higher or somewhat lower while approaching it.

<u>What questions to ask:</u>
1. Why does the circle disappear at a certain distance from your left eye?
2. Is the angle or the distance between the circle and the left eye more important in getting it to disappear?
3. Does it make any difference if we had other shapes on the paper?
4. What makes us see the circle although we're focussing on the triangle?
5. How does the human eye function?
6. Is the retina of our eye totally complete and continuous?

<u>How to explain it:</u>
    The retina of the human eye may be compared with the photographic film in a camera. The big difference is that in the human eye it is nerve cells that send electrical impulses to the brain when light rays fall on the retina. These nerves are bunched up and coming into the retina at a certain spot, called: **the blind spot,** and it is at this spot that very small images are undetectable (see Sketch on right).

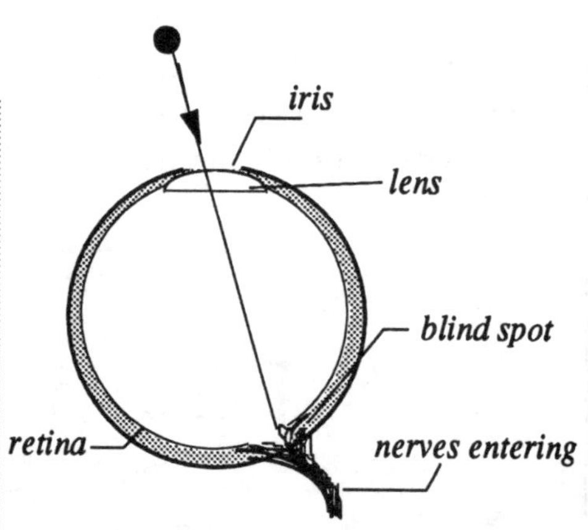

Top view of left eyeball

4.2.03. | **WHY DO WE NEED TWO EYES ?**

What you need: 1. A pencil & a piece of molding clay.

How to do it:

1. Place the pencil vertically in a piece of clay on the table top.
2. Let the students one by one try the following, while the rest of the class observes: Approach the pencil from the side about 3 - 4 meters away with one eye closed (cover the other eye with your hand). Hold the other hand stretched out and without hesitation, point down with the index finger and try to touch the pencil. Repeat again. Now repeat with both eyes open.

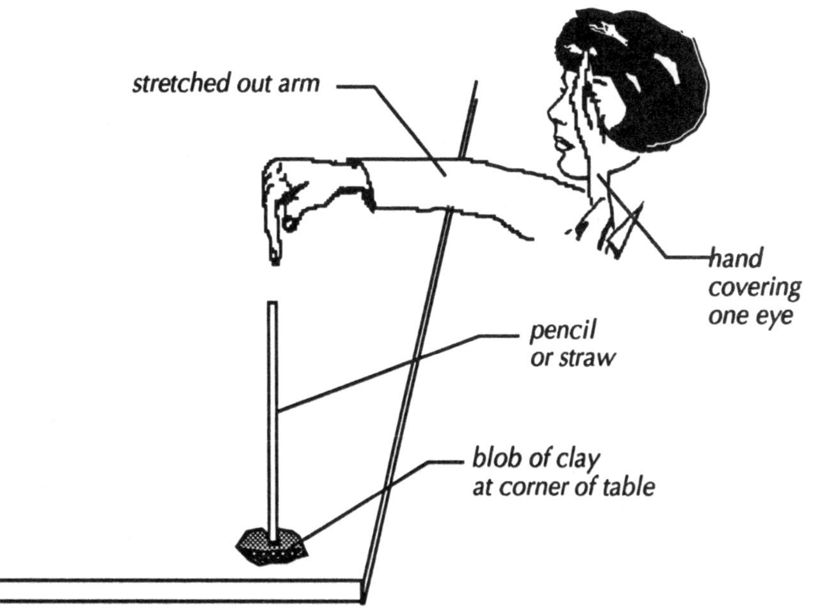

What to ask:

1. Why did most students miss touching the pencil end?
2. After a student tried to do the trick several times, why did he/she get better at touching the pencil?
3. Do you think it would be easier to do by approaching the pencil slowly?
4. What happens when the experiment is done with both eyes open?
5. What do you lack when just one eye is used?
6. Why must the pencil be touched without hesitation?
7. If we see differently with each eye, why don't we see two images when both eyes are open?

How to explain it:

Most people will not be able to touch the pencil on the first try, because they do not see with one eye, how far in front of them the pencil is located. **One cannot judge depth and distance as well with one eye** as with two. With one eye, one sees everything in the same plane (as in a picture). In other words, everything becomes two dimensional rather than three dimensional.

With a little practice one will get better at judging distances with only one eye.

4.2.04.   DO WE REALLY USE BOTH EYES ?

What you need: 1. Two cardboard cylinders (about 30 cm long)—the core of wrapping paper or paper towel suits fine.   2. Two different pages of a newspaper.

How to do it:
1. Place the two sheets of newspaper on the desk in front of you.
2. Hold the cardboard tubes in front of each of your eyes, such that you can look at one sheet with one eye and at the other sheet with the other eye.
3. Keep both eyes open at all times—can you read both pages at the same time? Which do you read more easily?

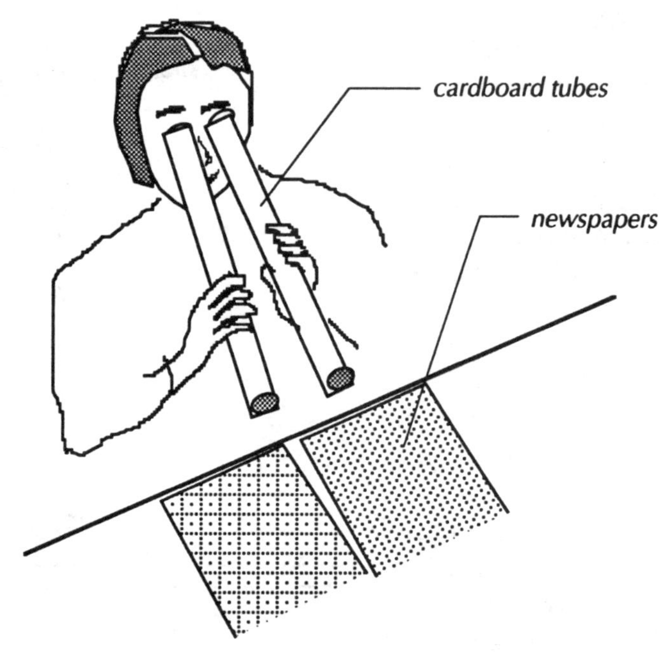

What to ask:
1. Which page do you read more readily?
2. Do you have to shut the other eye when reading one of the pages?
3. Why is it not possible to read both pages at the same time?
4. What does the preference of using one of your eyes indicate?
5. In what instances do we have to use just one eye?

How to explain it:
   We shall not be able to read both pages simultaneously, because we can only focus on one thing at a time. Although both eyes remain open, we can focus on one page or the other at will. It takes a simple act of will for one eye to see properly and for the other to stare blankly. It may thus be argued that we often make use of only one eye, even when we keep the other wide open. Evidently we do not have to cover one eye, either by holding a hand in front or closing the eye lid, when the use of only one eye is required, like when aiming a gun or looking through a microscope, or any other instrument requiring the use of only one eye.
   People, however, do prefer to use either the left or the right eye when they have the option, indicating that **there are left-sighted or right-sighted people,** just as there are left or right-handed people. Which group do you belong to?

4.2.05. | ARE YOU LEFT- OR RIGHT-SIGHTED ? (I)

What you need: 1. Two blank sheets of white paper.   2. A pencil.

How to do it:
1. Pierce a hole in the center of one sheet of paper with a pencil.
2. Draw a black dot in the center of the other sheet of paper the size of a penny, and place this about 40 cm in front of you on the table.
3. With both eyes open, hold the sheet with the hole between your face and the sheet with the dot, and move the first sheet about until the black dot can be seen through the hole.
4. While holding this sheet steady (while seeing the dot), close first your left eye and then your right. When does the dot disappear?

What to ask:
1. Does the dot disappear after closing your left or your right eye?
2. If the dot disappears when closing your left eye, are you left-sighted or right-sighted?
3. If the dot disappears when closing your right eye, are you left-sighted or right-sighted?
4. After determining that you are right-sighted, does it make any difference whether you are closing your left eye or not, in looking at the dot?

How to explain it:
For most people the dot will disappear when closing the right eye. This indicates that most people are **right-sighted.** It means that most people prefer to use their right eye over the left, if they are confronted with the option of using only one. In this case it means that those people can see the dot with both eyes open or with only the right eye open, but not with only the left eye. In other words, when both eyes are open, the left eye does not work. There is probably a connection between this phenomenon and the right-handedness of most people, although it would be hard to say which is the cause and which the effect.

4.2.06. ARE YOU LEFT- OR RIGHT-SIGHTED ? (II)

What you need: 1. A pencil or straw or ruler (or any straight edge).
2. A vertical straight edge of the room (inside/outside corners).

How to do it:
1. Hold the pencil (or straw, or any other straight edge) with an out-stretched arm vertically in front of you. (if no straight edge is available you can use your index finger and hold it vertically up).
2. Find a straight vertical edge (corner) of the room (or any other straight line) and align the pencil with it, holding <u>both</u> <u>eyes</u> <u>open.</u>
3. While keeping the pencil aligned with the straight edge of the room, close one eye at a time. Which eye do you close to make it look like the pencil jumps aside?

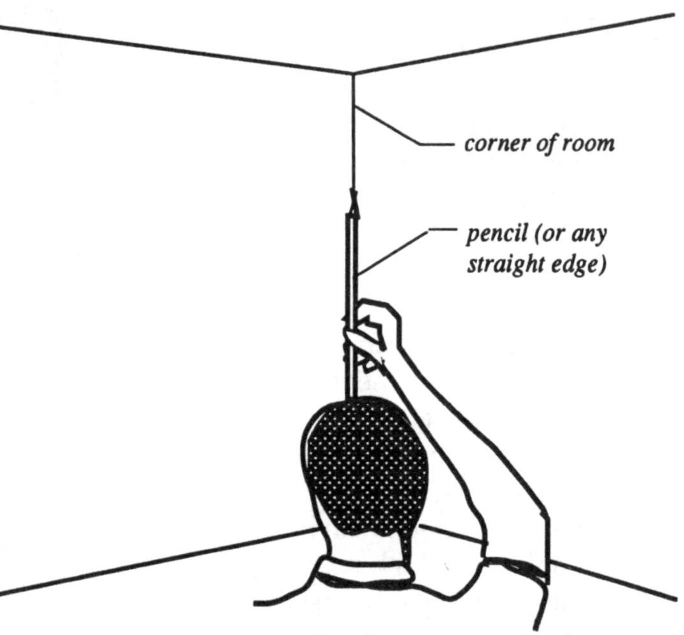

What questions to ask:
1. Does the pencil jump aside when closing your left or right eye?
2. If the pencil jumps aside when closing your right eye, are you using more your left or your right eye? In other words, is your left or right eye the dominant one?
3. If the pencil jumps aside when closing your left eye, are you left-sighted or right-sighted?
4. What makes the pencil jump aside when one of the eyes is closed?

How to explain it:
For the majority of people the pencil or whatever straight edge is aligned with the vertical line of the room, will jump aside when the right eye is closed. This indicates that most people are **right-sighted.** The reason for the pencil to jump aside, is that the right-sighted person uses his/her right eye more than the left. (the right eye is the **dominant** one). Thus the pencil is lined up in a straight line between the straight edge of the room and the right eye in the first place. When this eye is closed, the left eye still sees the pencil, but it is not in line with the edge of the room.

This is one of the easiest ways to tell whether a person is left-or right-sighted. If a pencil is not available, a forefinger held vertically may be used just the same. Left-handed people usually are left-sighted or in other words: their left eye is usually the dominant one, but there are always exceptions to the rule. These people that are the exceptions usually are **bi-dexterous**, meaning that they can use both hands with the same degree of skill or effectiveness.

4.2.07. | HOW DO WE PERCEIVE COLOR ? |

What you need: 1. A drawing of a heart-shaped figure (or other shape) having a yellow border, a green interior, and a small black dot in the center (on a double page size paper).
2. A plain double page size paper with a small black dot in the center of it.

How to do it:

1. Let the students (one by one) stare at the colored heart for twenty seconds without blinking their eyes, focussing on the black dot by holding the paper about 1 ft/ 30 cm from the face.
2. Immediately after, let them look at the plain piece of white paper, focussing on the black dot.
3. Ask the students to report what they see.

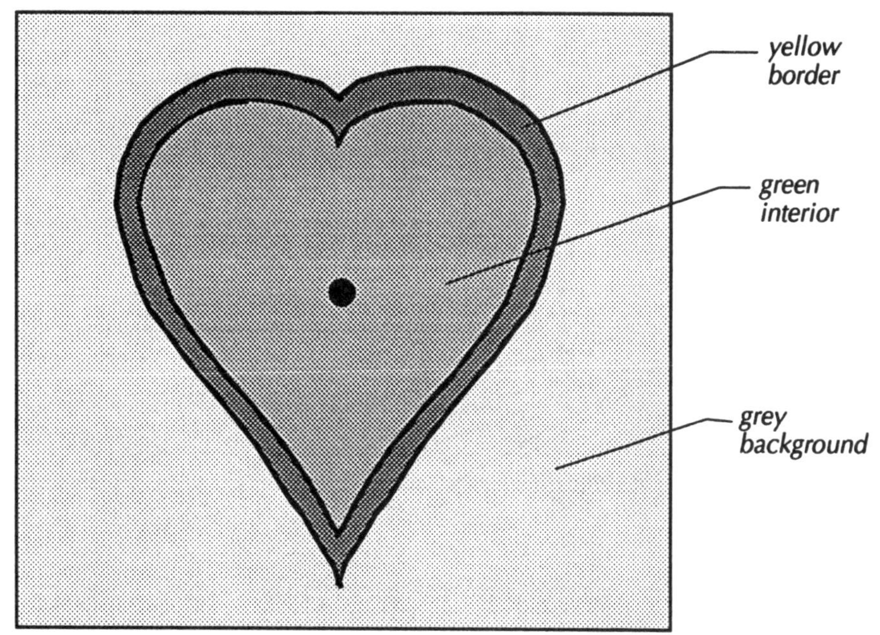

What to ask:

1. What color is the interior of the heart? The border?
2. What did you see when looking at the plain white paper?
3. What were the colors as compared to the first heart you viewed?
4. Why do you think the second heart was not the same as the first one?
5. Was there any heart drawn on the blank sheet of paper ?
6. What colors does the human brain pair together?

How to explain it:

There appears to be an unusual relationship between the eye and the brain in the human being. There is a relationship among four primary colors: red, green, blue, and yellow. As the retina absorbs light, **coders** farther back in the eye discriminate between the colors. The black-white coders can send a combined grey image, but one coder relays signals for green or red, and another coder for blue or yellow. These colors oppose each other and will not mix. **Steady exposure to any color tends to weaken the brain's response:** a bleaching effect makes the color fade or makes it grey. After staring at the yellow and green heart, a clear after-image was seen of a red heart bordered in blue. There is a temporary switching of signals to the brain. Since red and green share a single coding mechanism, as do yellow and blue, withdrawal of the color stimulus shuts down one part of the mechanism and triggers the other part for a moment or two.

The cones in the human eye pair the following colors: red-green, yellow-blue, and white-black.

4.2.08.   WHAT GIVES US THE ILLUSION ?

What you need: 1. Rulers (30 cm) & pictures or drawings of sketches below:

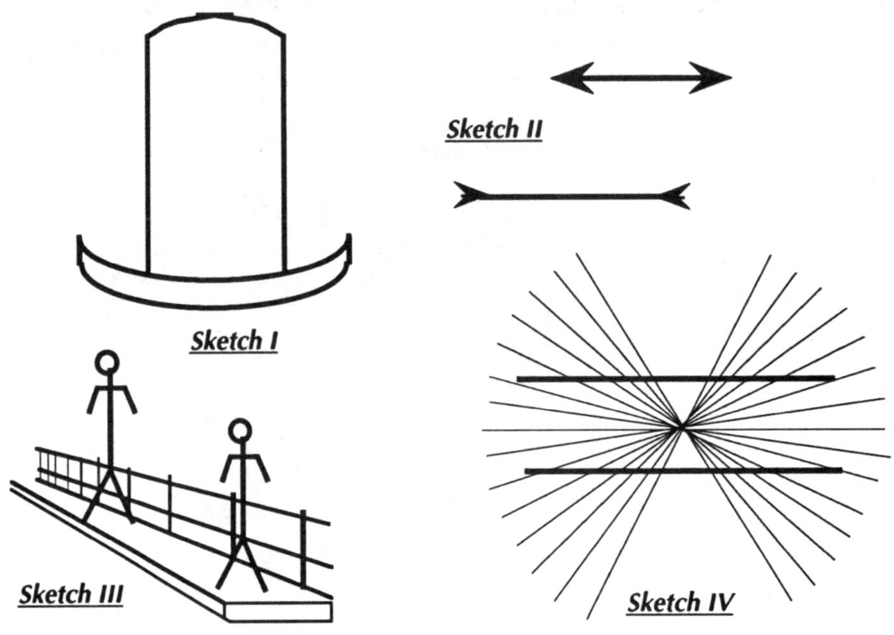

How to do it:
1. Let the students look at the pictures and ask the questions corresponding with each of the sketches.
2. Let the students measure or place the ruler along the lines.

What to ask:
1. How tall is the top hat compared to its width? (Sketch I)
2. Which of the two lines A or B is longer? (Sketch II)
3. Which of the two puppets is taller? Puppet A or puppet B ? (Sketch III)
4. Are the two lines A and B straight or curved? (Sketch IV)
5. What gave us the illusion in each of the sketches?

How to explain it:
In Sketch I of the top hat, people do not take the brim of the hat into account, and this is why the hat looks taller than it is wide. In the second sketch, people tend to look at the picture as a whole: line plus arrows at the end, and then line B looks longer. In picture III, the two puppets are seen in a three dimensional **environment**, where puppet B looks farther away than puppet A, and thus puppet B seems much taller. In Sketch IV it is the **environment** again that gives us the illusion that the two heavy drawn lines are curved.

By measuring each of the dimensions and by placing the ruler against the lines (for Sketch IV), we soon see that the dimensions of the hat, the two lines (in Sketch II), and the two puppets are exactly the same, and also that the two lines (in Sketch IV) are straight and running parallel to each other.

4.2.09. THE SWAYING CARDBOARD

What you need: 1. A white 5 x 8" cardboard or heavy paper.

Sketch A   *Sketch B*

How to do it:

1. Fold the paper card along the long axis in half.
2. Place the folded card directly in front of you on the table with the ridge pointing in your direction (See Sketch A).
3. Select a spot in the center of the fold and stare at it steadily with one eye shut (if you re right-sighted cover your left with your hand).
4. First you will see the card like in Sketch A. Continue staring at it until suddenly you see the card in its second position (like in Sketch B).
5. When you see the card in its second position (standing up), move your head slowly from side to side (still keeping one eye shut). Do you observe the card swaying back and forth?

What questions to ask:

1. How long did it take you to see the card in its standing position?
2. What did you see the card do when moving your head from side to side (with one eye shut)?
3. When seeing the card in its second position, are all the depth information and the way the light and shadow falls on paper and table, consistent?
4. What do you see when moving your head closer and farther from the card?

How to explain it:

This interesting **double illusion** demonstrates a phenomenon called **parallax.** When first seeing the folded card, it is lying down and you see it lying in the position as in Sketch A. All the information of the depth and how the light and shadows are falling, are consistent with our experience. When seeing it in the second position (like in Sketch B), the cues are all contradictory: the near point of the fold becomes the farthest point and the far corners of the paper become closest. Normally, when moving your head from side to side, near points will move more than far points of an object or scenery (try it with both eyes open). Seeing the card in its second position reverses these points and thus gives us an unusual distorted image. This is why we see the back side of the fold swaying so much!

4.2.10. SEE A HOLE IN YOUR HAND

What you need: 1. A cardboard cylinder (the core of paper towel or wrapping paper serves well) about 30 cm long (for each pair of students).

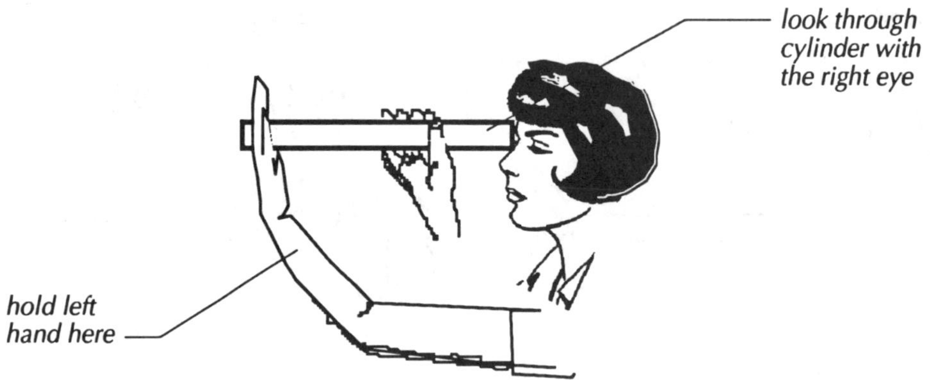

How to do it:

1. Pass the cardboard cylinders to the students (one for each pair or let two or three cylinders float around).
2. Demonstrate how to hold the cylinder and hand: Hold the cylinder with the right hand in front of the right eye, hold the left hand next to the cylinder close to the end of it, keep both eyes open and look (focus) at a distant point.
3. Now close each eye one at a time; open both eyes again. Do you see the hole in your left hand?

What to ask:

1. What did your left eye see (when the right was closed)?
2. What did your right eye see (when the left was closed)?
3. How did you come to see a hole in your left hand?
4. What do you have to do to see a hole in your right hand?
5. What can you infer from this, that the brain is doing with the vision or perception of each individual eye?

How to explain it:

By focussing at a distant point, the left eye in front of which the left hand is held, does register an image of the hand, but it is more the image of the right eye which is registered in the brain (because the eyes were focussed at a far point through the cylinder). When both eyes are open, it is the image of both eyes that are registered and combined by the brain. The left eye actually sees the left hand and the right eye sees the hole and **the brain combines both images.** This is the reason why a hole is seen in the left hand.

By switching the roles of the hands, it is possible to see a hole in the right hand.

4.2.11. THE FLOATING PIECE OF FINGER

What you need: 1. Yourself and your two index fingers.

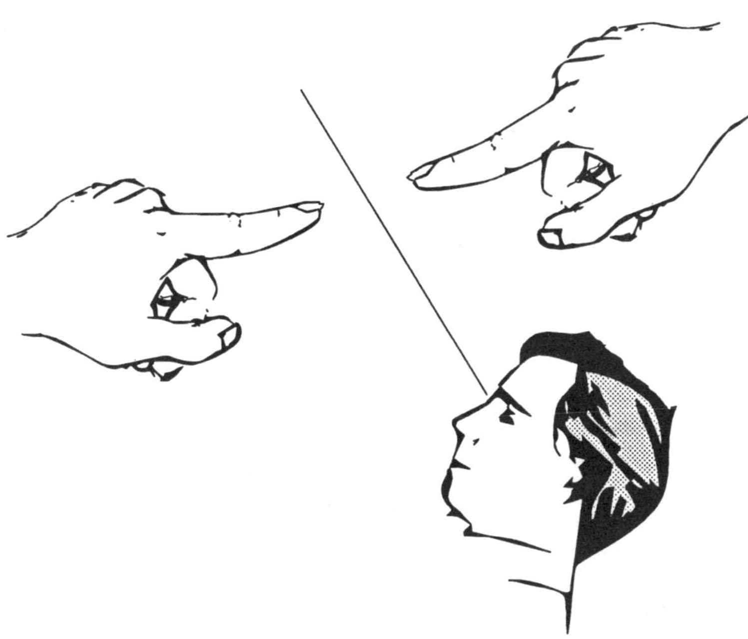

How to do it:
1. Hold your left and your right forefinger about 30 cm in front of you at the height of your eyes. Hold them horizontally about two-three cm apart (see Sketch on left).
2. Do not focus your eyes on the fingers but look over them and focus at a far point.
3. Wiggle the fingers slightly up and down. Can you see a double-nailed piece of finger floating in the air?! (Do not focus on your fingers!)

What questions to ask:
1. What made you see only a piece of each finger?
2. What happens if you focus your eyes on the fingers?
3. What happens if you look under the fingers and focus at a far point?
4. What do you see if you put your thumbs in the same position?
5. How can you check what your left eye is actually seeing?
6. Try closing one eye at a time; what do you see?

How to explain it:
Even though the eyes are focussed on a far point, we still see objects that are closer to the eyes. The **image** projected in the left eye and that projected in the right eye are both **combined** in our brain.

This is the reason why we see only a piece of the finger. Whatever image overlaps is seen more clearly. You would never be able to see the same floating piece of finger with one eye closed. We are not able to see depth with only one eye. It is like looking at a two dimensional picture, if we use only one eye.

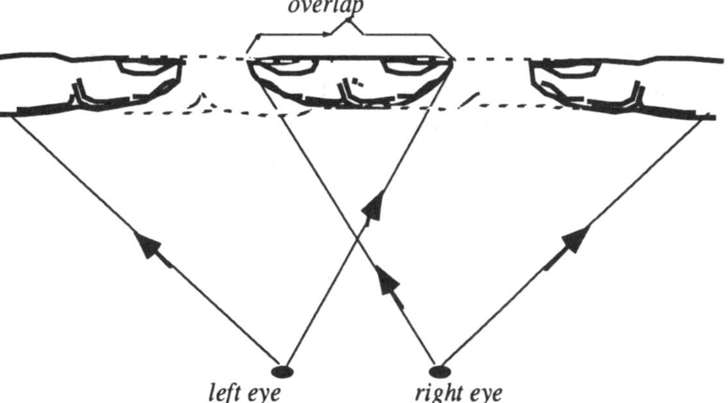

4.2.12. | THE HAND IS QUICKER THAN THE EYE |

What you need: 1. Two quarters or two other identical coins.

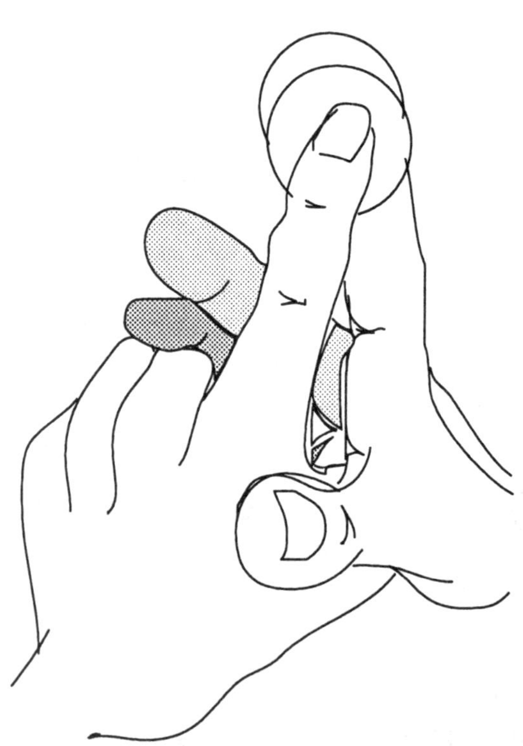

How to do it:
1. Place two coins between your two index fingers (conceal this from the students or audience, do not tell them how many coins you have).
2. Then rub the two coins back and forth as quickly as you can.
3. While rubbing, show them to the students (close up), and ask: "How many coins do you see?" Anticipated answer: "Three". If the answer is "Two", let them take a good look again at the moving coins.
4. After the majority of students (ideally all students) say that there are three coins between your fingers, go to the student that said "definitely three", and drop the two coins in his/her hand.

What questions to ask:
1. How many coins were actually moving?
2. Why do we see three coins when the coins are rubbed against each other?
3. What does it mean: the hand is quicker than the eye?
4. What do you see when the rubbing is considerably slowed down?
5. What are daily applications of this same principle?

How to explain it:
The human eye is a marvelous instrument. It operates on the same principle as the camera. The eye, however, is able to take two simultaneous pictures, one in black and white and the other in color. Cells called **rods** in the retina register black and white only; and other cells called **cones** register the different colors. The retinal cells are so sensitive that they can detect light as feeble as a 100-trillionth of a Watt. This is 1×10^{-11} Watt. The retina also has the capacity to **retain an image** a little shorter than **half a second,** this means that if the coins are moved faster than half a second per position, the eye still sees the coin at the original spot. Thus it sees three coins instead of two.

Other applications of this same principle are: motion pictures and animated cartoons, where sequential pictures are projected on the screen faster than half a second apart from each other and the eye interprets it as a continuous motion. Spokes of a moving wheel that sometimes are seen moving in reverse can only be seen when the vehicle in moving at a certain speed. In this case the eye actually retains the image of the spoke and since the next spoke is seen slightly behind this first one, it looks as if the wheel turns back.

4.2.13. | PUT THE BIRD IN THE CAGE

What you need: 1. A half dollar or dollar coin (a quarter is OK too).   2. Two large sewing needles.

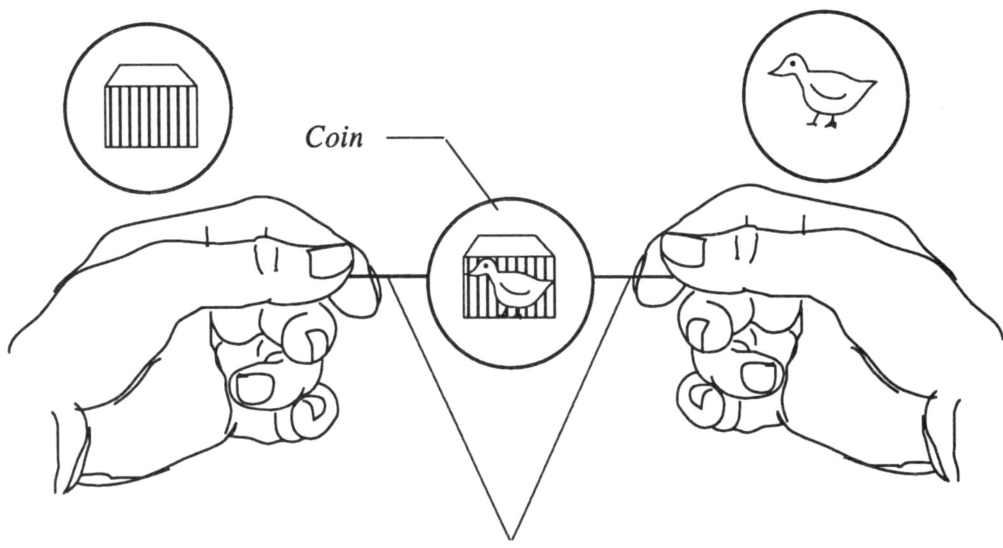

How to do it:

1. Cover both sides of the coin with tape that you can write on (masking or 3M magic tape) and trim the edges off.
2. First draw a cage on one side, then you draw a bird on the other side, making sure that you flip the coin upside down, and not sideways, before drawing the bird.
3. Show the drawings to the audience and ask them: "How can I put the bird in the cage?"
4. Find the diameter of the coin half way up the cage by cutting out a piece of paper the size of the coin and fold it in half.
5. Place the coin on top of the paper circle and pick the coin up with the two needles exactly along the diameter of the coin (see Sketch).
6. With the two needles bring the coin close to your mouth and blow against the upper half of the coin.
7. Repeat the blowing with short puffs. Observe the spinning coin: Bird in the cage!

What questions to ask:

1. Why do we place the two needles along a diameter of the coin?
2. What would happen if they were not placed along a diameter?
3. Around what point do spinning objects rotate?
4. What made us see the bird in the cage?
5. Did the bird actually go in the cage?

How to explain it:

All spinning objects spin around their center of gravity. This is why it is necessary to find a true diameter of the coin and place the two needles along this diameter. If the needles were not placed along a diameter, the spinning would not go very smoothly.

Seeing the bird in the cage is only an **illusion,** just like in moving pictures. The image of the cage or the bird is retained longer by our eyes than the time it takes for the coin to flip around. Thus both images are combined by our brain and we see the bird in the cage.

4.2.14. THE ELLIPTICAL PENDULUM SWING

What you need: 1. A thin string, a washer or nut (or other bob).
 2. A polaroid filter, or one lens of a pair of sun glasses, or semi-dark film negative.

How to do it:

1. Attach the washer or metal nut to the thin string to make a pendulum.

2. Let the pendulum swing from a fixed point (someone could hold the end of the pendulum with a steady hand).

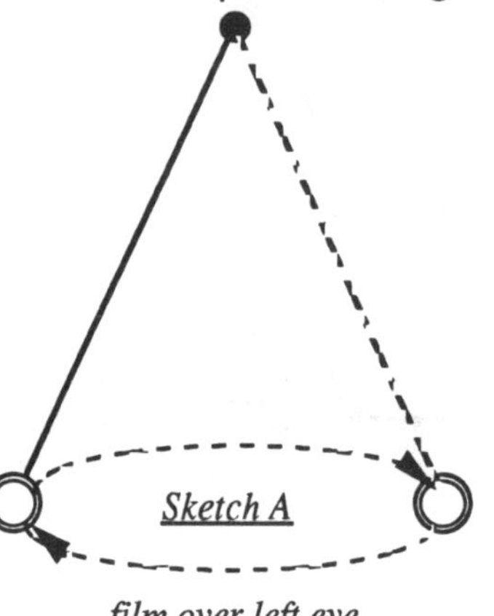

Sketch A

film over left eye

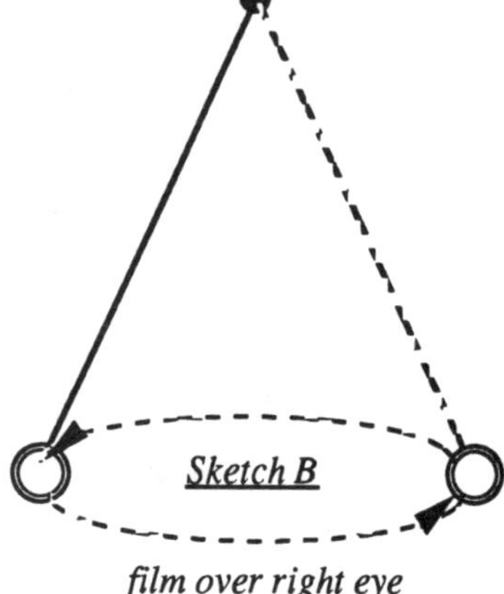

Sketch B

film over right eye

3. Look at the swinging washer with both eyes open, but cover your left eye with the polaroid filter (or other screens). What do you observe?

4. Now look at the swinging pendulum with the film over your right eye. What do you observe? In which direction is the bob swinging?

What questions to ask:

1. What is the reason for seeing the pendulum swinging in an eliptical path?

2. Would it be easier to see objects in a dark or in a bright room?

3. Would the eye perceive an object faster or slower if that object is brightly or less brightly illuminated?

4. If one eye is perceiving a moving object a fraction of a second slower, where would this eye actually see the object? In front or behind the actual position of the object?

How to explain it:

The eye that is covered by the darkened filter perceives the pendulum **a fraction of a second later** than the uncovered eye. This means that the left eye (say that this eye was covered with the filter) perceives the washer a little behind the actual position A (see Sketch below). Our brain interprets these two perceptions as coming from one object and combines the two images, thus we see it either a little farther or a little closer than the actual position.

The Sketch on the right illustrates the washer A moving to the right. The left eye (covered) perceives it at position B and the brain interprets it in position C. On the way back the brain interprets the position of the washer as being closer than the actual spot. Thus the motion is perceived as being eliptical.

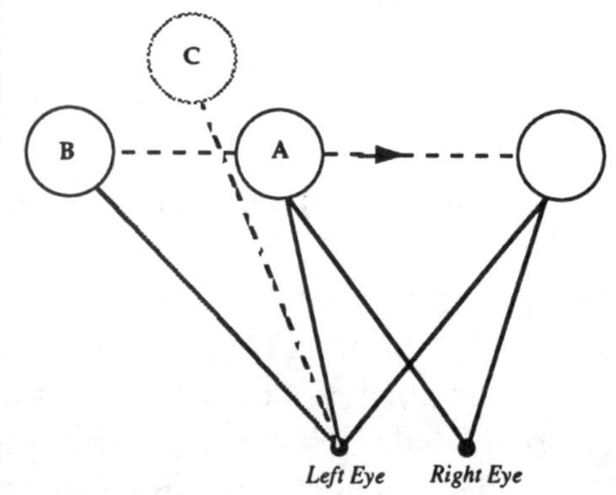

4.2.15. | HOW MANY POINTS ARE TOUCHING

<u>What you need</u>: 1. A bobby pin (hair pin) for each pair of students.

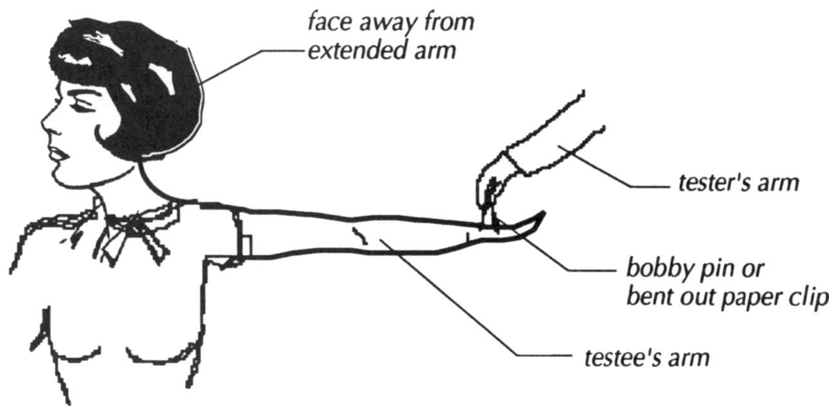

<u>How to do it:</u>

1. Let students work in pairs: one be the testee and the other the tester. The latter holding the pin.
2. Instruct the tester to spread the two tips of the bobby pin and place it on the testee's finger, hand, and arm, moving slowly up the arm starting from the middle finger, touching the skin randomly either with one or two tips.
3. The testee should look away from the touched arm (see sketch) and tell the tester how many tips he/she feels (without looking at the pin).
4. The tester should keep track of the number of mistakes made at the different parts of the arm. (Let tester and testee change roles).

<u>What to ask:</u>

1. Where was it easier to tell how many tips were touching?
2. Why did we make more mistakes the higher up the arm the pin was touched to the skin?
3. What makes our touch sense more sensitive?
4. Where would a cut in the skin hurt most?

<u>How to explain it:</u>

This activity demonstrates the distribution of the **nerve endings** in the human body, in our case from the fingertips to the shoulder. It is increasingly difficult to tell without looking how many tips are actually touching the skin, the higher up the arm the pin is placed. This is caused by the fact that there are many more nerve endings in the fingertips and hands compared to the upper arm. It would therefore be more painful to experience a cut in the skin closer to the fingertips than higher up the arm. For this same reason, our sense of touch is most sensitive at our fingertips compared to other parts of the body.

4.2.16.   IS THE WATER WARM OR COLD ?

What you need: 1. Three beakers (400 ml) or large cups.

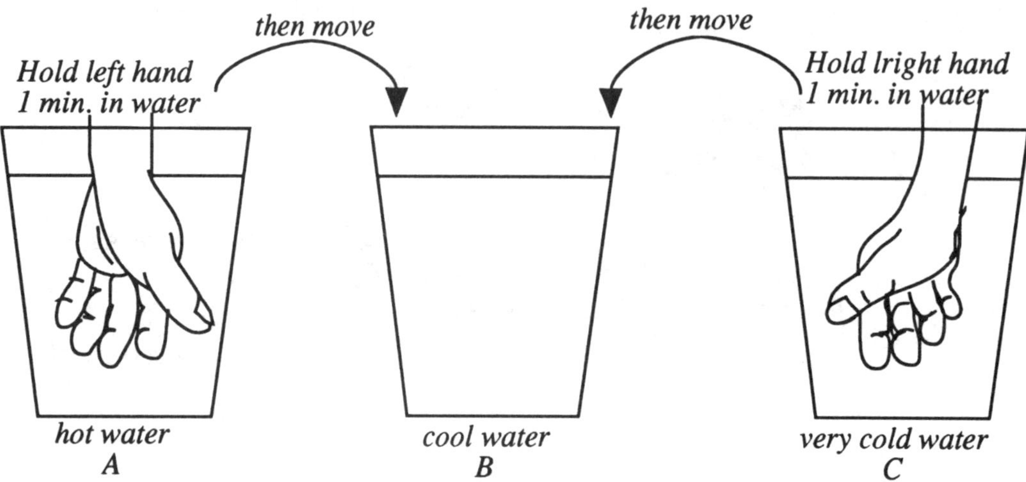

How to do it:

1. Fill a beaker (A) three quarters full with hot water (about 50°C) and another beaker (C) three quarters full with very cold water (about 5°C) (add a few ice cubes if water is not cold enough in the summer time).
2. Fill the third beaker (B) with regular water of room temperature (20°C) and place it in between the hot and the cold water beaker.
3. Immerse your left hand in the hot water and hold it under water for one minute, then move it in the center beaker. Does the water feel warm?
4. Immerse your right hand in the cold water and hold it under water for one minute, then move it in the center beaker. Does the water feel warm or cold?

What questions to ask:

1. Did the water in the center beaker (B) change in temperature?
2. Why did the water in beaker B feel warm with one hand and cold with the other?
3. What gives us the sensation of warm or cold in our body?
4. What are other similar situations where this same principle applies?

How to explain it:

When the left hand is placed in the hot water, the **nerve endings** in our nervous system send a message to our brain, telling it that it feels warm. Over time, however, the nerve endings are dulled and the hand adapts itself to the warm sensation. When it is placed into the water of room temperature, there is a difference of about 30°C lower, and the water suddenly feels cold to this hand. The same interpretation can be put forth for the hand immersed in the cold water. This time it's only going the opposite way: from cold to warm. Similar situations are encountered when we take a shower with very cold hands and adjusting the water temperature with the touch of our cold hands; placing our body under the shower, the water still feels cold because to the hand the shower water already felt very warm.

Someone used to living in a home with 18°C room temperature might find on visiting a friend's house with a 22°C room temperature quite warm, and of course visa versa.

4.2.17. | **CATCH THE DOLLAR BILL**

What you need: 1. A crisp dollar bill (any denomination) (ideally: one of each denomination up to $20).

How to do it:
1. Hold your forefinger and thumb of your left hand about 5 cm (2 inches) apart, place the dollar bill about half way in between the two fingers.
2. Let the dollar bill fall and show the audience that you can easily catch it between your two fingers.
3. Ask the members of the audience to hold their forefinger and thumb in the same position to try to catch the dollar bill.
4. Hold the dollar bill between their fingers. The rule is that they may not go down with their hand, and the gap between the fingers should not be smaller than 5 cm.
5. Vary the time between placing the dollar bill between their fingers and the moment that you let go of the bill. Tell the audience that whoever catches the bill may keep it. Continue doing it with larger denominations, but remember to vary the dropping time, otherwise the bill might get caught by anticipation.

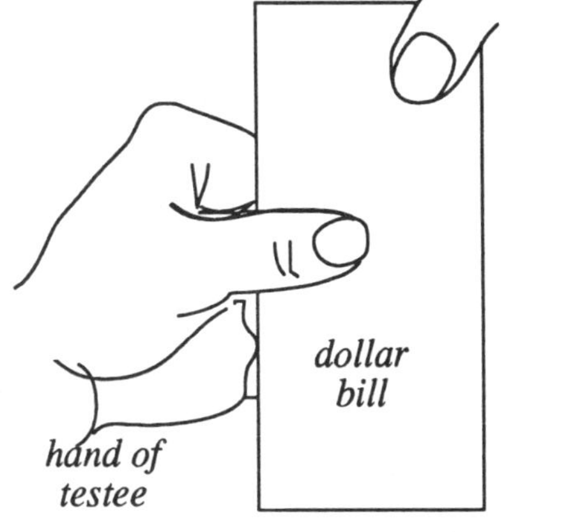

What questions to ask:
1. Why can you catch the dollar bill so easily when you drop it yourself?
2. Why do you need to vary the time between placing the dollar bill and the releasing of it?
3. What is the reason that the bill can never be caught?
4. What makes muscles contract?
5. How did the finger muscles get the order to catch the bill?
6. Where did the initial stimulus came from?

How to explain it:
The initial **stimulus** to catch the dollar bill was from seeing the falling of it. The eyes got the stimulus, they sent a signal (**response**) to the brain and the brain in turn sent an **electrical impulse** to the finger muscles to contract and catch the bill. But all that took a **longer time** than the falling of the dollar bill, although both take only a fraction of a second.

If you place the dollar bill between your left fingers and let it fall with your right hand, it is very easy to catch. This is because the stimulus or signal was already in your brain, and all it had to do is send the signal to the muscle of your fingers. The difference in doing it to yourself or having someone else do it to you, is the time it took for the light to travel from the dollar bill to the eye, and for the eye to send a signal to the brain.

The bill can only be caught by someone who anticipates the falling and actually start closing the fingers before the bill falls. This is the reason why we should vary the time between placing the bill between the fingers and the actual releasing of it.

4.2.18. | HOW FAST CAN YOU REACT ?

What you need: 1. A meter stick (15 meter sticks for a class of 30).

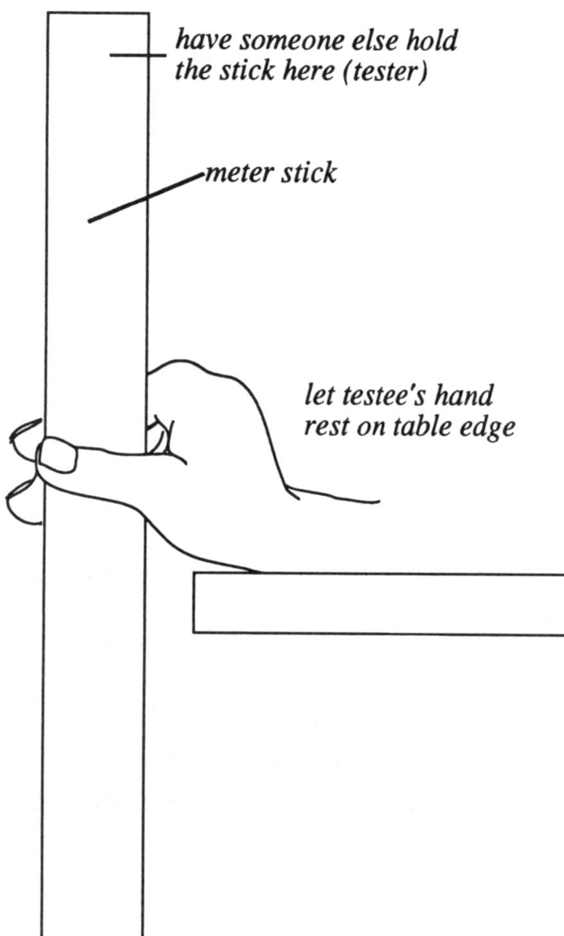

How to do it:

1. Place your hand over the edge of a table and leave about 3 cm opening between the thumb and forefingers.
2. Let your tester (someone else) hold the meter stick vertically from the top end of the stick.
3. Read off the spot where your thumb is on the meter stick before the tester drops the stick (start at an even number f.i. 10 or 20).
4. Let the tester drop the stick. The moment that you see the stick drop catch the stick immediately and do another reading of where your thumb ended up. Record the difference (distance in drop).
5. To test the other stimuli: <u>Hearing:</u> close your eyes and let tester say: "Now" exactly at the moment that he/she drops the stick. Testing for <u>tactile (touch):</u> close your eyes and let tester touch your other hand exactly at the moment that he/she let go of the stick.

What questions to ask:

1. How do girls compare to boys in reaction time?
2. In doing the above comparison, what variables have to be controlled?
3. How do the different stimuli compare in your reaction time?
4. What is the actual falling time for the distance in the drop?
5. What are all the variables involved in this experiment?

How to explain it:

The different variables in this experiment are: the time of release of the ruler, the way the ruler is held, the time between saying "now" and releasing the ruler (when testing audio), etc. The **manipulated variable** is the **stimulus:** through the eyes, hearing, and touch (or sex: when comparing boys versus girls). The **responding variable** is the falling time or distance on the meter stick. As the distance in free fall is: $d = 1/2\ gt^2$, the time derived from that is t = square root of **2d** over **g** , where **g** is the gravity acceleration.

All other variables, like: distance between fingers, the place where the tester holds the stick, catching with two fingers or with the whole hand, the accuracy of saying "now" and simultaneously letting go of the stick, etc. have to be held constant, in other words they have to be done the same way.

4.2.19. THE STIMULUS-RESPONSE ACTION

What you need: 1. A wooden rod or block (a ruler would do).
    2. Paper and pencil for each student.

How to do it:
1. Stand in the rear of the class behind the students (some place where the student cannot see you).
2. Instruct the students to make a tally mark every time you say "write."
3. Strike a desk with the wooden rod and simultaneously say "write" for a total of 20 times with 2 second intervals.
4. Continue striking the desk, but stop saying "write" until everyone has stopped writing.
5. Have the students count their tally marks.
6. Tell the students that you said "write" exactly 20 times.

What to ask:
1. Why did most students make more than 20 tallies?
2. What was the stimulus and what was the response?
3. Why did I pair the strikes with the saying of "write"?
4. What was the purpose of the tap with the wooden rod?
5. Why did everyone stop writing eventually?

How to explain it:
When a **response-eliciting stimulus** (the saying of "write") is paired with a **neutral stimulus** (taps with the wooden rod), the student becomes **conditioned to the neutral stimulus,** so that the neutral stimulus will elicit the response, even when it is presented alone. The conditioning will become extinguished after a certain time, if the presentation of the **conditioned stimulus** (tapping noise) is continued without pairing with the **unconditioned stimulus** (the saying of "write").

This conditioning of the human being to neutral stimuli occurs in daily life for example, when the preparation of food or a meal is constantly accompanied by the clatter of kitchen utensils; hearing only the clatter will stimulate the feeling of hunger or might even stimulate the production of saliva in the salivary glands (drooling).

4.2.20. | HOW FAST DOES YOUR HEART BEAT ?

What you need: 1. Low step stools.   2. A stopwatch or a watch with a second

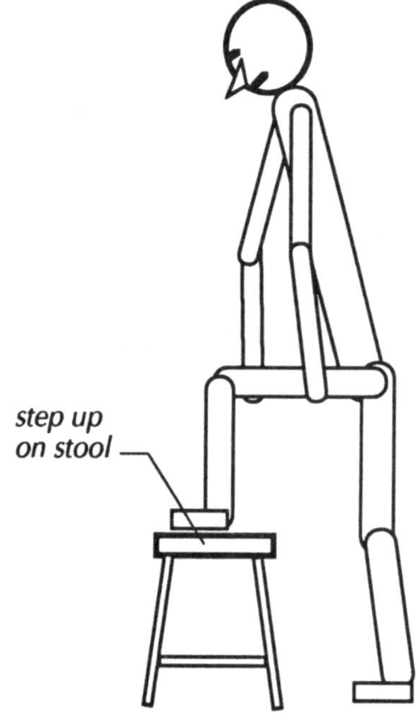

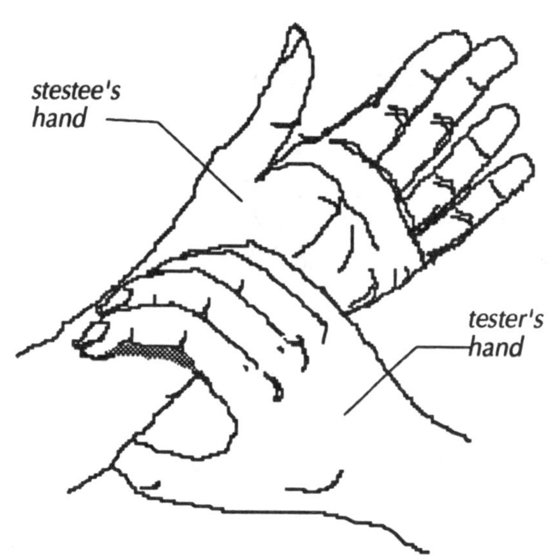

How to do it:

1. Let students work in pairs: one being the tester and the other the testee.
2. Instruct the tester to count the number of heartbeats of the testee by placing his fingers on the testee's wrist, and counting the pulses within 10 seconds (and multiply this by 6).
3. Now let the testee step up and down the stool 10 times, and immediately after that let the tester take the testee's pulse (repeat point 2).
4. The number of steps up and down the stool can now be increased to 20 and 30, immediately followed by a pulse count.
5. Let the tester and testee change roles.

What to ask:

1. What was the pulse before the steps up and down the stool?
2. What was the pulse after 10, 20, and 30 steps up and down the stool?
3. What made the heart beat faster after the exercise?
4. What did the muscles need, to do all that work?

How to explain it:

The pulse or heartbeat is caused by the **blood pressure impact on the arteries** as the heart muscles contract. It can be felt by placing a finger on the **radial artery** at the wrist. By doing vigorous exercise, like the steps up and down the stool, the leg muscles need more oxygen and thus more blood to carry this oxygen. The heartbeat thus automatically speeds up to pump more blood to the working muscles, from around 72 beats per minute to more than 120. Within three minutes this speeded pulse should return to the normal 72. With increasing number of steps up on the stool, an increase in the pulse should be observed in the testee. Some deviations from these values may be normal for certain individuals.

4.2.21. | WHICH CONTAINS MORE CARBONDIOXIDE ?

What you need: 1. Two Erlenmeyer flasks.   2. Saturated lime water (calcium hydroxide).
    3. Two sets of glass tubing in 2-hole stoppers (see sketch).

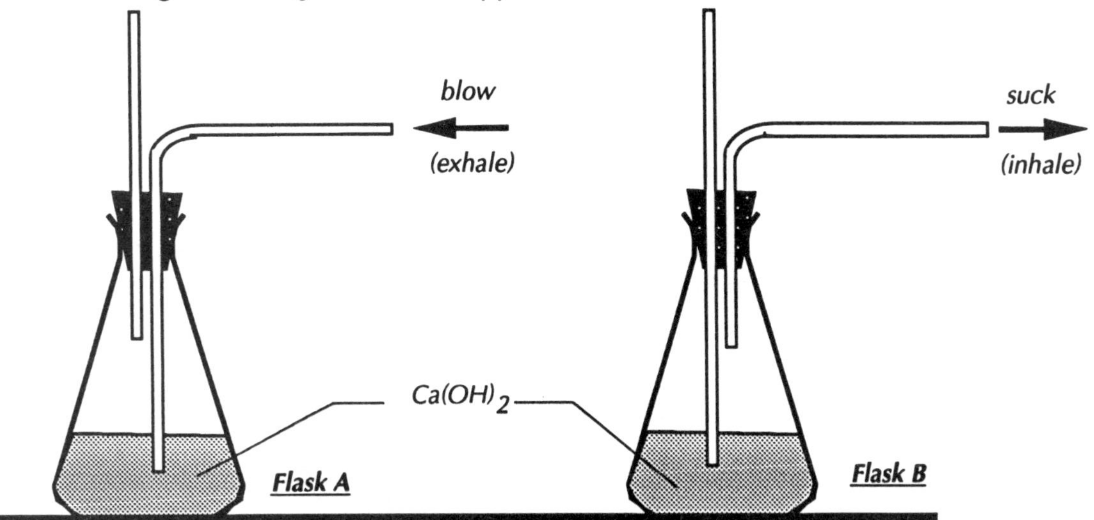

How to do it:
1. Make a saturated solution of calcium hydroxide by adding small amounts of the solid powder, a little at a time to warm water while stirring, until no more solid dissolves.
2. Fill each flask half full with the lime water and cover them with the two-hole rubber stoppers (and fitted glass tubing).
3. Suck (inhale) through the tube which ends above the water surface in flask B and exhale through the tube that extends into the liquid in flask A (see sketch).

What to ask:
1. Which of the two flasks is getting milky first?
2. What gas is actually led through the liquid in flask A?
3. What does it indicate if the liquid turns milky?
4. What chemical reaction is taking place?
5. Will flask B eventually also turn milky if the sucking is continued?
6. What will the liquid in flask A eventually do if the blowing through it is continued?
7. Which of the two, inhaled or exhaled air, contains more carbondioxide?

How to explain it:
    By sucking through the tube in flask B, **atmospheric air** is led through the saturated calcium hydroxide solution (this air is the same as the air we inhale). By blowing through the tube in flask A, exhaled air is led through the liquid, which turns milky sooner than that in flask B. This indicates that there is more CO_2 present in exhaled air. The reaction is as follows:
    $CO_2 + Ca(OH)_2 \rightarrow CaCO_3 + H_2O$, in which the calcium carbonate **precipitates out** and makes the solution appear milky. The CO_2 in the air will eventually turn the liquid in flask B also milky, and when the blowing is continued in flask A, the reaction: $CO_2 + CaCO_3 + H_2O \rightarrow Ca(HCO_3)_2$ will turn the liquid clear again (Ca-bicarbonate is **soluble** in water).

4.2.22. | MEASURE THE CAPACITY OF YOUR LUNGS

What you need: 1. A large glass jar (pickle gallon jar).
2. An unused aquarium (or other transparent container).
3. A length of rubber tubing (about 40 cm long).
4. A 1 liter measuring cylinder (or other liter container).

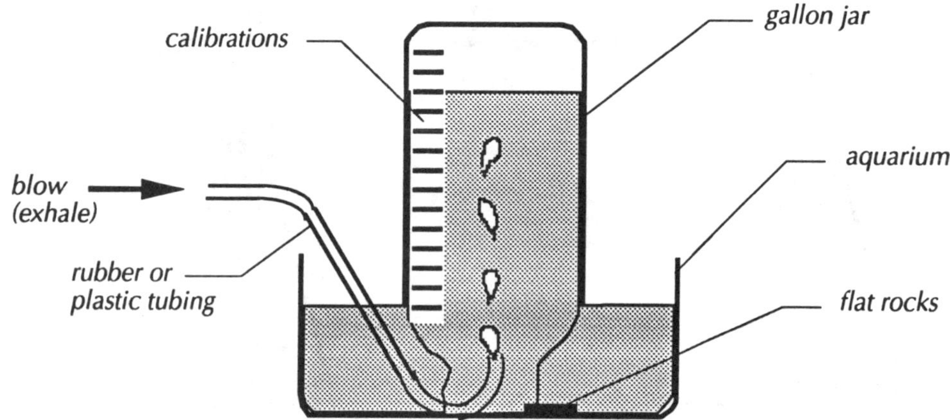

How to do it:
1. Calibrate the large jar by filling it with water, liter by liter, and marking the water level with a marker or masking tape after every liter increase.
2. Fill the gallon jar completely full and the aquarium 3/4 full with water.
3. Place three flat stones of the same thickness (or other heavy objects) on the bottom of the aquarium, and invert the gallon jar into the water-filled aquarium so that it rests on the three stones.
4. Insert one end of the rubber tubing under the mouth of the gallon jar and let the other end hang over the rim of the aquarium.
5. Let one student hold the inverted jar steady and another student, whose lung capacity is to be measured, exhale through the tube after inhaling as deeply as he/she can.
6. Measure volume of exhaled air in jar.

What to ask:
1. What made the water stay up in the inverted jar?
2. Why did the student have to inhale as deeply as possible before exhaling?
3. In exhaling, what does the student have to do in order to obtain a true measure of his/her total lung capacity?

How to explain it:
By inhaling as deeply as we can, we are actually filling our lungs full with air. When we blow all the air out through the tube and catch this exhaled air in the jar above the water, the **volume** or **capacity** of our lungs can thus be measured. This we do by reading off the volume of air in the jar. The larger this lung capacity, the more it is indicating that the individual involved is enjoying better health.

4.2.23. HOW DO WE BREATHE ?

What you need 1. A large bottle with a rather narrow neck.
2. A one-hole stopper (fitting in the bottle neck).
3. A Y-glass tube (or straight tube).   4. Hot oil & an ice cube.
5. Two small balloons & rubber bands.   5. One large balloon (or beach ball balloon).

How to do it:

1. Cut the bottom of the large jar out, by covering it with a cm thick layer of hot oil, and touching the outside of the jar with an ice cube exactly at the surface of the oil (the jar will crack and the bottom will drop out). Smoothen the edge by fil-ing or firing.
2. Tie the two small balloons to the ends of the Y-tube with the rubber bands. Hold the long end of the Y-tube through the jar neck and insert the one-hole stopper over the Y-tube and in the jar neck (see Sketch).
3. Cut the large balloon in half, stretch it over the open bottom of the jar and tie or tape it around airtight (it represents the diaphragm).
4. Pull the center of the diaphragm up and down and observe the balloons expand and collapse.

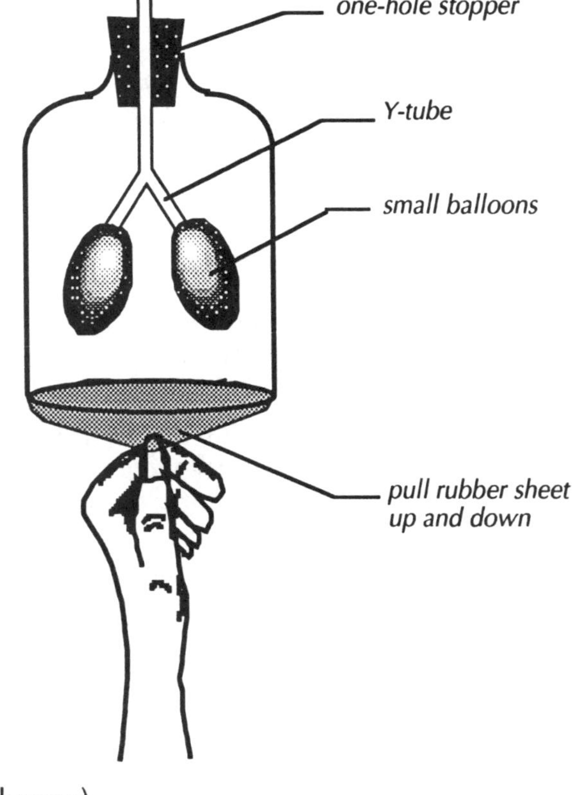

What to ask:

1. What does the Y-tube represent in this demonstration?
2. What do the small balloons and the stretched rubber sheet represent?
3. What made the small balloons expand?
4. What stage of the breathing could the expanded state of the balloons be compared with?
5. How different would it be to compare the glass jar with our chest cavity?
   How are they the same?

How to explain it:

The Y-tube in our demonstration represents the **bronchial tube**, the small balloons, **the lungs**, the rubber sheet, the **diaphragm**, and the jar, the **chest cavity**. This latter one, however, can be expanded or contracted through the flexible rib joints, which is not the case with the glass jar. By pulling the rubber sheet down, the pressure inside the jar decreases, thus sucking the air from outside into the small balloons (inhaling).

4.2.24. THE UNCONTROLLABLE FOOT

What you need: 1. A blank sheet of paper.   2. A pencil or pen.

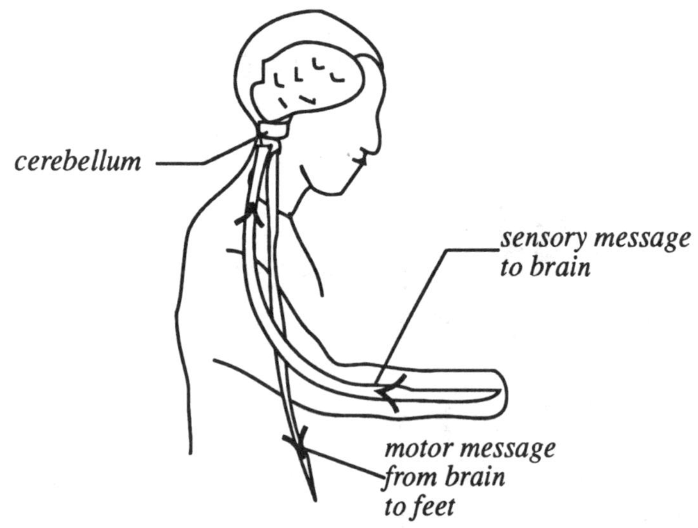

How to do it:
1. Stand up next to the table with the paper in front of you.
2. Hold the pen or pencil in your right hand (if you're right-handed).
3. If you're right-handed, let your right foot rotate in a clockwise direction (describing small circles on the floor with you right foot).
4. Try to keep your foot rotating while simultaneously writing a large *number 6* on the paper. Observe what your foot is doing!

5. Now repeat point 3 and 4 but with your left foot making the clockwise motions. Try drawing the six now with your right foot making counterclockwise motions. Then do the same with your left foot. What are your findings now?

What questions to ask:
1. Which of the four combinations was easiest to carry out?
2. When moving your foot in a clockwise direction, was it easier to do it with your left or right foot (while drawing the 6)?
3. Did you try to write the sixes with your left hand (if you re right-handed) while rotating your foot?
4. What makes it so difficult to rotate your right foot (when right-handed) or your left foot (when left-handed) in a clockwise direction?
5. What part of the human body controls the muscles?

How to explain it:
   The **cerebellum,** located in the base of the **cerebrum** of the human brain, controls the coordination of voluntary movements. The cerebellum is important for such activities as walking, dancing, playing ball, or even for such routine tasks as tying a shoelace or writing a figure 6.
   Doctors have known for generations that nerve fibres from the right side of the body cross over in the brain stem to the left side of the brain. Similarly, nerve fibres from the left side of the body cross over to the right side of the brain. In other words, the whole **left side of the body is controlled by the right half of the brain**, and similarly, the whole right side of the body is controlled by the left half of the brain.
   By writing a figure 6 with the right hand, the left half of the brain has instructed the right hand to make a counter-clockwise motion. The right foot could easily make the same counter-clockwise motion, but the opposite movement requires a special effort. With practice both movements can be achieved. Drummers have no trouble performing this, because their muscles have been trained to do independent movements!

4.2.25.   THE KICKING FROG LEG

What you need: 1. A freshly cut frog leg.   2. A copper (bronze) bolt, and two nuts.
   3. An iron stand or frame (galvanized iron strip will do).

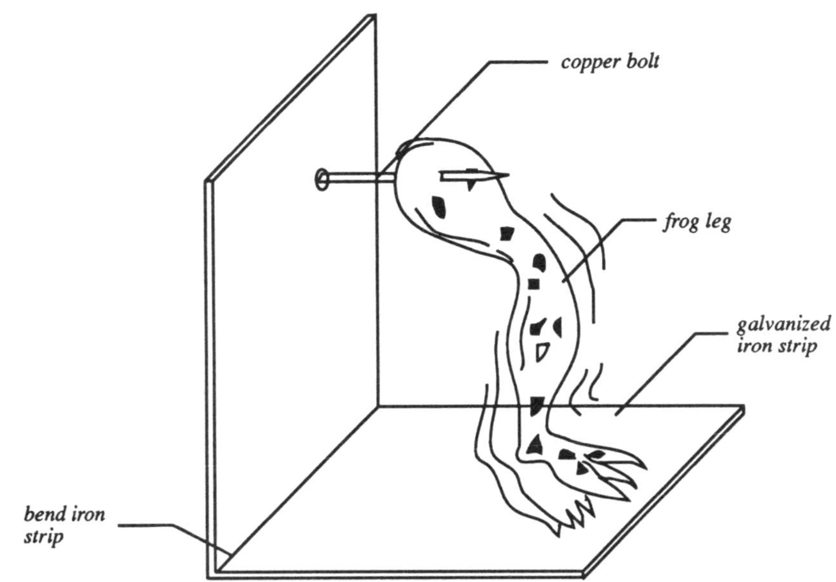

How to do it:
1. Drill a hole in one end of the iron strip and attach the copper bolt tightly to this strip with the two nuts.
2. Bend the iron strip at a spot in such a way that when the frog leg is hung from the bolt, the foot will touch the lower part of the iron strip.
3. Pierce the frog leg through the copper bolt in the thigh and let the leg droop down. Observe the spasms!

What questions to ask:
   1. What made the frog leg go into spasms?
   2. What would be necessary for a muscle to contract?
   3. Would an iron bolt do the same thing for the frog leg?
   4. Would this demonstration work if you use an iron bolt in a copper stand?
   5. What other metals would give this same contraction in the frog leg?
   6. How can we find out which metals give the potential difference?
   7. How can we compare this with human muscles and their contraction?
   8. Why is it dangerous to grab a live wire on purpose?
   9. How should we test a live wire with our hands?

How to explain it:
   There is a potential difference between copper and iron. At the moment that the frog leg touches the iron frame, while it is suspended from the copper bolt, an electrical current (a flow of electrons) is running through the muscle and makes it to contract. After a while the leg muscles relax and the leg droops down, the foot touches the iron and the leg goes into spasms again.
   The further apart two metals are in the **electromotive series,** the larger the **potential difference,** in other words the larger the difference in tendencies to go into solution or give off electrons. Other combinations are: copper and zinc, magnesium and copper, iron and silver, etc.
   Human muscles are also controlled by electric currents. It is therefore very dangerous to deliberately grab a "live" wire, as you may not be able to let go of the wire. In order to test a live wire, use the back of your fingers or hand and touch it for a short moment. If the wire sends electricity through your muscle, your fingers will be moving away from the wire rather than clamping around it tighter, if you touch it with the inside of your hand.

4.2.26. | ARE WOMEN MORE AGILE THEN MEN ?

What you need: 1. A plastic cup (or any other object of the same dimensions).

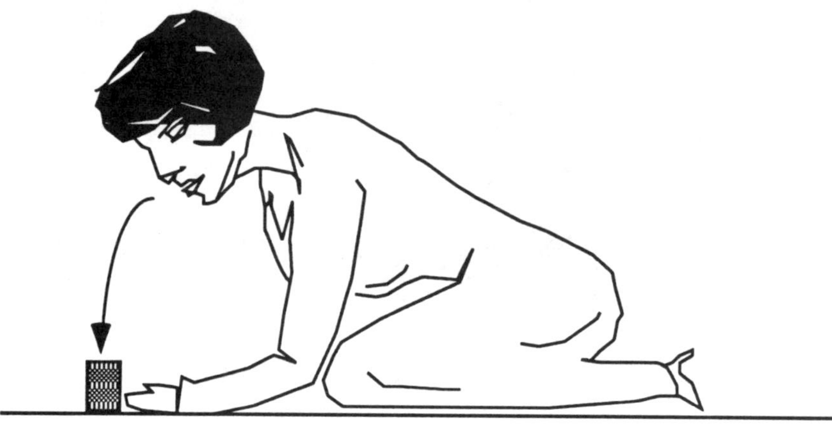

How to do it:

1. Let one of the ladies in the audience sit on her knees and lower leg (like in a kneeling position) on the floor.
2. Place a cup (or other small object) in front of her the length of her forearm away from her knees on the floor (see Sketch).
3. With arms behind her back, let her bend forward and knock the object over with her nose! Men cannot do this without falling forward! Woman can!

What questions to ask:

1. Why can women knock the object over without falling, but not men?
2. Are women actually more agile and lean compared to men?
3. What is different in the basic skeleton built of women compared to men?
4. Where is the center of gravity of women located?
5. Where is the center of gravity of men located as compared to women?
6. How would the exceptions of women that cannot do the trick be built?
7. How would the exceptions of men that can do the trick be built?

How to explain it:

In general, women's body structure is such that they have a lower **center of gravity,** because of their wider hips and heavier bone structure in the lower abdomen part of the skeleton as compared to men's structure. Similarly we can say for men in general, that they have wider shoulders as compared to women. This makes the center of gravity of men's bodies higher than women's.

This lower center of gravity is the main cause for women to be able to bend forward in the kneeling position without falling over forward. When men try to bend over they will fall forward because of the location of their center of gravity. When this center of gravity passes beyond the knees, their bodies will topple forward. (See also Event 3.1.10 for a similar activity).

4.2.27. TURN A CRACKER INTO SUGAR

What you need: 1. Unsalted and unsweetened crackers.

How to do it:
1. Distribute one cracker to each student.
2. Let them chew the cracker without swallowing it for a minute or two.
3. Ask them the following questions.

What to ask:
1. How did the cracker taste in the beginning of the chewing?
2. How did the cracker taste at the end of the chewing period?
3. What was the cracker mixed with in your mouth?
4. What does your saliva do to the cracker while chewing?
5. What agent in your saliva breaks down the starch molecules?
6. Are sugar molecules smaller or larger than starch molecules?
7. Why is it better to chew food a little longer before swallowing?
8. What would happen if we almost did not chew our food?

How to explain it:
The cracker that was put in the mouth consists of **carbohydrates (starch)**. This being unsalted and unsweetened, will taste quite bland in the beginning of the chewing period. While the cracker is being pulverized into the pulpy mass called a **bolus,** the **digestive juice** of the **salivary glands** begins a breakdown of the carbohydrates. An **enzyme (amylase)** in saliva splits the molecules of starch into smaller molecules of sugar. This enzyme cannot break down starch particles that are still enclosed in their natural cellulose envelopes, therefore starches should be cooked before eating.

Another purpose of the saliva is to provide moisture needed by the taste buds. Saliva also has a cleansing action on the teeth. It washes away food particles that otherwise might provide a home for bacteria.

By chewing food a little longer before swallowing, it gets mixed more thoroughly with our saliva, providing an opportunity for the enzymes to break down the large starch molecules, and by the time the food reaches the stomach and the intestines, it can be easily digested.

4.2.28. | TRY DRINKING WHILE STANDING ON YOUR HEAD

What you need: 1. A glass of drinking water or juice.
2. A bent drinking tube or flexible drinking straw.

How to do it:
1. Ask one of the students to assist you, to put the straw in your mouth as soon as you are ready to drink.
2. Try to stand on your head against the wall (if this is not possible, bend from your waist down until your head touches the floor).
3. Have your student assistant bring you the glass and straw so that you can drink from it.
4. Suck through the straw while your body (or upper body) is upside down, and empty the whole glass.

What to ask:
1. Is gravity needed to make fluids come down the esophagus?
2. Can we drink water while we stand on our head?
3. How does food go down the esophagus into the stomach?
4. Is the esophagus like a glass or rubber tube going in the stomach?
5. Why did the fluid not flow out of the mouth when drinking upside down?
6. What is the muscle action called, which pushes food into the stomach?

How to explain it:
After food is swallowed, it enters the **esophagus**. Here it is pushed along by **muscle action**, which is called **peristalsis**. This movement of the food is carried out by **involuntary muscles**. There are two layers of muscles: the inner layer forms a series of circles around the tube, and the outer layer is longitudinal. When the inner layer contracts, the tube becomes smaller at that point, and when they relax the longitudinal muscles contract. This **alternate contraction and relaxation** of the two sets of muscles push the food along the tube in **peristaltic waves**. This is the reason why food, whether it is in solid or liquid form, may be swallowed with the body positioned in any direction. Gravity has little or no influence on this process.

DEVELOP THE ATTITUDE OF

"I CARE FOR YOU"

AND YOU'LL HAVE

THE WHOLE WORLD WITH YOU !!

YOU'LL BECOME WHAT YOU THINK ABOUT !

- WILLIAM JAMES

AN AUDIENCE OF TEACHERS
AT ONE OF DR. TIK LIEM'S SEMINARS

HOW ABOUT BECOMING

THE **GREATEST** SCIENCE MOTIVATOR AND DEMONSTRATOR IN THE WORLD.. ...OR WHATEVER YOU FANCY!

WITH GOD'S HELP!

INDEX

TEACHERS' TEXTS & GUIDES

Item No.	Description	Price
B1	**"INVITATIONS TO SCIENCE INQUIRY" - 2ND ED**................. By: Tik L. Liem, 507 pages, with over 400 discrepant events to motivate students in learning science. Illustrations and explanations on every page!	$45.00
B2	**"INVITATIONS TO SCIENCE INQUIRY" - Supplement to First and Second Edition,** by Tik L. Liem;.175 pages with 150 additional (50 over 2nd Ed) discrepant events and explanations!.......	22.50
B3	**"IDEA BANK COLLATION"** - Irwin Talesnick, Editor .... around 500 pages containing over 600 teaching ideas in science!	35.00
B4	**"TURNING KIDS ON TO SCIENCE in the home"** by Dr. Tik L. Liem.(Set of 4bks)	70.00
B4-1	**Book 1 - Our Environment,** Easy to follow recipes for doing over 150 activities in Air, Weather,Matter & Chemicals in our environment with very simple materials found in and around the home................	22.50
B4-2	**Book 2 - Energy,** Easy to follow recipes for doing over 120 science activities on Heat, Magnetism, Electricity, Light & Sound with very simple materials found in and around the home........................	20.00
B4-3	**Book 3 - Forces & Motion,** Easy to follow recipes for doing over 100 science activities on Forces and Motion on Earth and in Space with very simple materials found in and around the home........................	17.50
B4-4	**Book 4 - Living Things,** Easy to follow recipes for doing over 50 science activities on Plants and Human Biology, carried out with very simple materials found in and around the house........................	15.00

Discount of 10%: for orders of 16 texts or more.

VIDEO TAPES: with TIK L. LIEM

Item No:	Description	Price
V1	**"DISCREPANT EVENTS - THE SCIENCE MOTIVATOR"**........... The cognitive dissonance rationale, plus an assortment of discrepant events presented to teachers. Observe how they bring students to the teachable moment! (60 min)	40.00
V2	**"TURNING KIDS ON TO SCIENCE"**... A different assortment of discrepant events demonstrated to science students of grade 7. Observe what excitement they create in the students! (60 min)	40.00
V3	**"GETTING HOOKED ON SCIENCE" with DR. TIK L LIEM** .... A demonstration of exciting, discrepant events to an assembly of grade 7-9 students. Observe how powerful a tool the discrepant event is for the science teacher ! (45 min)	45.00
V4	**"EXCITING SCIENCE FOR THE HOME" with DR. TIK L LIEM**...... A different variety of discrepant events demonstrated to middle school students and their parents, stressing the use of very simple materials that can be found in and around the home (60 min).	45.00

Discount of $5.00 per video: for orders of 2 or more videos.

MATERIALS & CHEMICALS

Materials :

ITEM	EVENT NO.	CHAPTER	TITLE/DESCRIPTION	PRICE
			ASSORTMENT A	
MA1	2.3.10	Magnetism	Make a Needle Compass	$ 2.00
MA2	2.7.16	Sound	The Twirling Bugle	5.00
MA3	3.1.20	Forces	The Floating Belt Hanger	4.00
MA4	3.1.38	Forces	The Yip-yip Stick	4.00
MA5	3.1.47	Space Science	The Immovable Penny	2.00
MA6	3.2.07	Space Science	Get the Chalk in the Bottle	3.00
MA		**Package of all six items under A above**..........		**17.50**
			ASSORTMENT B	
MB1	1.1.25	Air	Inverted Paper Bag Balance	2.50
MB2	1.4.08	Char of Matter	Invisible Flame Extinguisher	2.00
MB3	2.1.04	Energy	The Tin Can Bazooka (Lighter Fuel Cannon)	8.00
MB4	2.3.04	Magnetism	The Floating Discs	4.00
MB5	3.1.05	Forces	The Balancing Pins	5.50
MB6	3.2.23	Space Science	The Funny Marbles	3.00
MB		**Package of all six items under B above**..........		**22.50**
			ASSORTMENT C	
MC1	1.1.34	Air	The Ballancing Balloons	3.00
MC2	1.2.03	Flowing Air	The Funnel and the Ball	4.00
MC3	1.4.20	Char of Matter	Pour Water along a String	3.00
MC4	2.2.11	Heat	Which is the Warmer Color?	6.00
MC5-1	1.4.30	Char of Matter	Thread Circle in the Soap Film)	
MC5-2	1.4.31	Char of Matter	The Strong Soap Film)3 in one packet	16.00
MC5-3	2.1.08	Energy	Fission and Fusion)	
MC		**Package of all materials under C above**..........		**27.50**
			ASSORTMENT D	
MD1	1.5.05	Chemistry	Draw With Fire	3.00
MD2	1.5.08	Chemistry	Turn Water into Wine, Milk, and Beer!	7.50
MD3		Chemistry	Magic Flash Paper (5 different colors)	12.50
MD4	1.5.25	Chemistry	The Vanishing Water (Polyacrylate)	7.50
MD5	1.5.19	Chemistry	The Complete Combustion (Lycopodium powder)	5.00
MD6-1	2.4.04	Static Electricity	Separate the Pepper from the Salt)2 in one packet	3.00
MD6-2	2.4.05	Static Electricity	The Pepper Reproduction)	
MD7	3.1.24	Forces	The Magic Strip of Newspaper	5.00
MD		**Package of all materials under D above**..........		**37.50**

TEESHIRTS:

T1: YIP-YIP STICK , Size XL, L, or M 16.00/ea

T1+   plus Yip-Yip Stick........... 18.00/set

T2: FLOATING BELT-HANGER, Size XL, L, or M..................................... 16.00/ea

T2+   plus Belt-hanger.............. 18.00/set

Please add shipping & handling costs: For orders of:

up to $100........... add 15% of total order.

$101 to $300........... add 10% of total order.

$301 and up............ add 5% of total order.

Make checks out to: **Science Inquiry Enterprises**
14358 Village View Lane
Chino Hills, CA 91709
Tel/Fax: 714-590-4618

MINIMUM ORDER OF MATERIALS: $17.50

ORDER FORM   92/93

Amount	Item No.	Description	Unit Price	Total Price
		TEXTS		
_______	B1	Invitations to Science Inquiry - 2nd Ed, by Liem	$45.00	_______
_______	B2	Invitations to Science Inquiry - Suppl 1st & 2nd Ed, Liem	22.50	_______
_______	B3	Idea Bank Collation, I. Talesnick, Editor	35.00	_______
_______	B4	Turning Kids On To Science in the home-Dr. Tik L. Liem		
		Set of four books	70.00	_______
_______	B4-1	TKO Book 1 - Our Environment (186 pages)	22.50	_______
_______	B4-2	TKO Book 2 - Energy (168 pages)	20.00	_______
_______	B4-3	TKO Book 3 - Forces & Motion (126 pages)	17.50	_______
_______	B4-4	TKO Book 4 - Living Things (71 pages)	15.00	_______
		VIDEOS		
_______	V1	Discrepant Events - The Science Motivator	40.00	_______
_______	V2	Turning Kids On To Science	40.00	_______
_______	V3	Getting Hooked On Science, with Dr. Tik L. Liem	45.00	_______
_______	V4	Exciting Science For The Home, with Dr. Tik L. Liem	45.00	_______
		MATERIALS		
_______	MA	Assortment A	17.50	_______
_______	MB	Assortment B	22.50	_______
_______	MC	Assortment C	27.50	_______
_______	MD	Assortment D	37.50	_______
_______		Individual material sets, item Nos._____________		_______
		TEESHIRTS		
_______	T1	Yip-Yip Stick Teeshirt , Size XL, L, M	16.00	_______
_______	T1 +	Teeshirt plus the Yip-Yip Stick	18.00	_______
_______	T2	Floating Belt-Hanger Teeshirt , Size XL, L, M	16.00	_______
_______	T2 +	Teeshirt plus the Belt-hanger	18.00	_______

Sub-total: ________

Add shipping & handling accordingly (15%, 10% or 5% of sub-total): ________
For California residents add 8.25% Tax (of sub-total) ________

Please make checks payable to: **Science Inquiry Enterprises**
14358 Village View Lane
Chino Hills, CA 91709
Tel/Fax: 714-590-4618

Grand total: ________

Send Invoice to:

____
(Name)

____
(Address)

____
(City)   (State)   (Zip)

____
(Tel)

Send Materials to:

____
(Name)

____
(Address)

____
(City)   (State)   (Zip)

____
(Tel)

NOTES